C000233760

ATLANT &
THE
POWER SYSTEM
OF THE GODS

ADVENTURES UNLIMITED PRESS

The **Lost Science Series**:
- VIMANA AIRCRAFT OF ANCIENT INDIA & ATLANTIS
- TECHNOLOGY OF THE GODS
- THE GIZA DEATH STAR
- THE GIZA DEATH STAR DEPLOYED
- VIMANA AIRCRAFT OF ANCIENT INDIA & ATLANTIS
- ATLANTIS & THE POWER SYSTEM OF THE GODS
- LOST CONTINENTS & THE HOLLOW EARTH

The **Lost Cities Series**:
- LOST CITIES OF ATLANTIS, ANCIENT EUROPE
  & THE MEDITERRANEAN
- LOST CITIES OF NORTH & CENTRAL AMERICA
- LOST CITIES & ANCIENT MYSTERIES OF SOUTH AMERICA
- LOST CITIES OF ANCIENT LEMURIA & THE PACIFIC
- LOST CITIES & ANCIENT MYSTERIES OF AFRICA & ARABIA
- LOST CITIES OF CHINA, CENTRAL ASIA & INDIA
- ANCIENT TONGA & THE LOST CITY OF MÜ'A
- ANCIENT MICRONESIA & THE LOST CITY OF NAN MADOL

The **Atlantis Reprint Series:**
- THE HISTORY OF ATLANTIS by Lewis Spence (1926)
- ATLANTIS IN SPAIN by Elena Whishaw (1929)
- RIDDLE OF THE PACIFIC by John MacMillan Brown (1924)
- THE SHADOW OF ATLANTIS by Col. A. Braghine (1940)
- ATLANTIS MOTHER OF EMPIRES by R. Stacy-Judd (1939)
- SECRET CITIES OF OLD SOUTH AMERICA by Harold Wilkins (1952)
- MYSTERIES OF ANCIENT SOUTH AMERICA by Harold Wilkins (1949)

The **New Science Series**:
- THE TIME TRAVEL HANDBOOK
- QUEST FOR ZERO-POINT ENERGY
- THE FREE ENERGY DEVICE HANDBOOK
- THE FANTASTIC INVENTIONS OF NIKOLA TESLA
- THE ANTI-GRAVITY HANDBOOK
- ANTI-GRAVITY & THE WORLD GRID
- ANTI-GRAVITY & THE UNIFIED FIELD
- ETHER TECHNOLOGY
- THE ENERGY GRID
- THE BRIDGE TO INFINITY
- THE HARMONIC CONQUEST OF SPACE
- UFOS & ANTI-GRAVITY: Piece for a Jig-Saw
- THE COSMIC MATRIX: Piece for a Jig-Saw, Part II
- THE AT FACTOR: Piece for a Jig-Saw, Part II
- THE TESLA PAPERS

ATLANTIS
&
THE
POWER SYSTEM
OF THE GODS

MERCURY VORTEX GENERATORS
& THE POWER SYSTEM OF ATLANTIS

BY
DAVID HATCHER CHILDRESS
& BILL CLENDENON

ATLANTIS & THE POWER SYSTEM OF THE GODS

Copyright 2000
David Hatcher Childress
and Bill D. Clendenon

First Printing
April 2002

ISBN 0-932813-96-8

Mercury: UFO Messenger of the Gods
Copyright 1990 by Bill Clendenon

Printed in the United States of America

All Rights Reserved

Published by
Adventures Unlimited Press
One Adventure Place
Kempton, Illinois 60946 USA
auphq@frontiernet.net
www.wexclub.com/aup
www.adventuresunlimitedpress.com
www.adventuresunlimited.co.nl

10 9 8 7 6 5 4 3 2

Acknowledgements

Many thanks to Christopher Dunn, Stephen Mehler, Moray B. King, Stan Deyo, John Anthony West, NOVA, the Association of Research and Enlightenment, the Unarius Society, the Giza Pyramid Research Center, Dr. Howard John Zitko, The Lemurian Fellowship, NOVA, Mark Lehner, Kathy Collins, Harry Osoff, Jennifer Bolm and many others.

The illustration on page 274 is courtesy of Stephen Mehler

TABLE OF CONTENTS

Part 1
Foreword

by
David Hatcher Childress

Neville's Conclusion:

*Research is a straight line from the tangent
of a well-known assumption to
the center of a foregone conclusion.*

The Strange Case of Bill Clendenon

Bill, I hardly knew ye. He would call me on the phone sometimes, or send me stuff in the mail. He sent me his first book, the *ASP Survival Manual* (the ASP stood for Adventure Survival Publications), with some clippings about him and his adventures.

When my book on ancient Indian vimanas came out in 1991, Bill contacted me and sent me a paperback book that had been published the year before. He hired the new publisher of Ray Palmer's old magazine, *Search*, to bring his book out in paperback. In 1990 Bill's book *Mercury: UFO Messenger of the Gods* had been released to an unwary reading public. Bill had loved my book, and thought I would appreciate his.

I did. It was an odd-ball assortment of rants, diagrams, photocopied letters (and even envelopes) that was fascinating to browse through. I was especially interested in Bill's concept of the mercury-plasma-gyros that he thought he had decoded from the *Vimanika Shastra* text. Most of

i

Bill at a UFO conference, circa 1990.

Bill's mercury
vortex diagram.

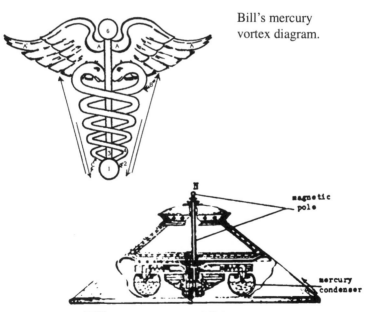

Bill's mercury powered flying saucer-vimana.

his book is reprinted here in full, as it appeared in the second edition.

Not many people ever really saw that first book, except for a few that Bill had sent out, and a handful sold through *Search* magazine and a few mail order UFO outlets. The book featured a photo of a jaunty-looking Bill with his explorer's hat on. With his bushy eyebrows and cockeyed expression, he looked to me like a character out of a TV western.

Bill would go to UFO conferences and tell his stories. He had known George Adamski and was a witness himself to a UFO that he claims looked much like the Adamski craft. Indeed, other simi-

Bill Clendenon.

lar craft, such as the 1964 Utah photo, do appear to have been seen and photographed by witnesses. Bill unfortunately did not have a camera with him at the time he saw the saucer-shaped craft, with three small pods beneath the disk.

Later, the next day, Bill received a strange visitor... but we'll let him tell that story. Needless to say, Bill was a believer in UFOs, but his "spin" on the UFO situation was different from that of most "contactees" and UFO researchers. He believed that flying saucers were from planet Earth, had been manufactured thousands of years ago, and were powered by a type of electrified mercury gas—a plasma—that was engineered into a gyroscopic vortex action much like the toroid action of a smoke ring.

Bill was described by Judith M. Statezny, the publisher of *Search*, as "a cross between Indiana Jones and a leprechaun, a sweet man with only one purpose—to inform the public of his findings and research. We became fast friends and decided to collaborate on a book."

Because Bill had read my *Lost Cities* books, was a believer in Atlantis, and a student of vimanas and the ancient Hindu epics, he would call me and we would talk about the ancient legends, the messenger god Mercury, and Bill's many theories on mercury gyroscopes and related topics.

In 1993, Bill released his book again with a local printer in Biloxi,

Mississippi. He was living there on social security, renting a room at the back of an elderly lady's house. Bill came and went, walking down to the store, sending some books out to customers and waiting for the next UFO convention. Then one day his landlady called me and said that Bill had passed away. She had found him lying on his bed with his clothes on, arms outstretched. Apparently a heart attack had struck one afternoon.

Bill is gone, but his work lives on. I promised Bill that I would keep publishing on vimanas, Atlantis, mercury gyroscopes, and the technology behind the phenomena known as UFOs.

My own investigations showed me that Bill was not a "nut" at all, but rather, very advanced in his thinking on the entire enigma of UFOs and ancient astronauts. There were advanced ancient civilizations. They had flight. They apparently used the liquid metal mercury as a power source...

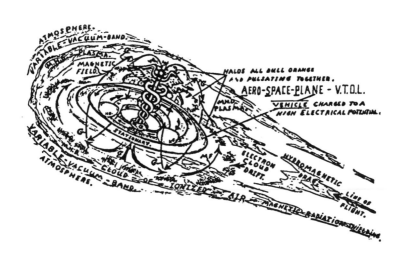

Mercury, the messenger god who can fly.

INTRODUCTION
BOOK I

(1) ALIEN: "Owing political allegiance to another country or government; foreign"

(2) EXTRATERRESTRIAL: "Originating, located, or occurring outside Earth or its atmosphere"

(3) UFO: "Unidentified flying object (this term is used at times in place of flying saucer)"

UFOs and flying saucers are real. Mercury, the Herald or Messenger of the gods, has entered into our midst via the UFO and, in doing so, introduced new beginnings and the confusion that inevitably precedes new beginnings. A Herald, of course, should state his business plainly, but Hermes is mercurial and elusive in nature making it difficult to decode or translate the highest mysteries of the Gods and one of the highest mysteries of the Gods is UFOs (flying saucers).

This two-part book represents over 40 years of my interest, research, and UFO encounters in the field. The author is convinced that there is an international conspiracy to keep the existence of UFOs from the people of the world. In the beginning, I was all for cooperating with government authorities and professional researchers within the scientific community regarding UFOs. For, like most people, I was taught to cooperate and trust persons in positions of authority and higher education. But, they fumbled the ball, wasted public funds, and added insult to injury by discrediting, by fair means or foul, all those who tried, by whatever means available to them, to inform the public that flying saucers or UFOs were indeed an international reality.

In the second part of this book, I hope to awaken and communicate with professionals in the aerospace field as well as the general public, in hopes that we can pool our experiences, knowledge, and resources to force the truth of

UFO reality out in the open. One way is to duplicate, at least in model form, the propulsion system of the type of UFO or aerospace plane (ASP) to which the Adamski UFO Scout Ship of 1952 points. This in turn may open the door to things and events of far greater magnitude that will benefit all of mankind in times to come.

For those readers wanting to know the general history of UFOs or more about George Adamski, the man, I refer you to the Recommended Reading at the back of this book.

There is an old saying, ''One picture is worth a thousand words.'' Some of the UFO photographs herein are worth more words than this book can contain. Regarding the value of photographs and the analysis of same for information, the American taxpayer shells out a great deal of money for the government's so-called black budgets to purchase all kinds of sophisticated surveillance cameras for spy planes and satellites. Yet, the public is told by government and so-called authorities within the scientific field that photos of UFOs are not valuable scientific evidence in helping to prove that UFOs or flying saucers are real. All those video cameras out there may yet prove them wrong.

In 1986 U.S. Senator John Glenn (D-Ohio), a congressional watchdog, ex-military pilot, and former astronaut, became irate and complained that the Pentagon was stonewalling the existence of super secret stealth aircraft, when any kid with $9.95 in his jeans could buy a plastic scale model of stealth aircraft at a toy store. The stealth model was closely reproduced by the Tester Corporation's model designer John Andrews by using public information and common sense, and the man had not even seen a stealth aircraft! Whereas, the findings of my years of UFO research efforts were gleaned from copies of the UFO photographs within the pages of this book, backed by my personal sightings and encounters with the actual UFOs or flying saucers in the field. The U.S. government, after 40 years of stonewalling and double talk, continues to deny the existence of UFOs or flying saucers to this day. I am saying, as a tax paying, voting American citizen and UFO witness, that I have had it with government incompetence. I am, I will, and I can do something about it. *One vote does make a difference!*

As an American citizen, I regard our government as a trustee when conducting the business of our country. As a trustee, the government is bound by laws, which includes the Constitution of the United States, to conduct itself as a government of the people, by the people, and for the people. I believe that the only secrets our government is justified in keeping from the citizens, for a limited time only, are the ones which would endanger the country. I believe that anybody who handles anyone else's money, including monies in the government's black budgets, or monies spent on covert or intelligence operations, should give a complete accounting of every penny. I also believe that no contract is binding that is signed in secret by a few individuals to bind the whole populace.

In the matter of UFOs and flying saucers, I feel that the government is not on the side of the citizens of The United States. The government, a trustee, denies the years of experience of the American citizens' own eyes, investigations, logic, reasoning, and firm convictions based on positive identification by experienced observers. According to some UFO reports, it is rumored that hard scientific evidence is in the possession of the U.S. government. However, we need the help of Congress to verify this one way or the other. With the help of Congress, the public may well uncover proof to our satisfaction that some UFOs and/or flying saucers are aerospace craft produced and operated by a mysterious race of human beings; human beings representing an unknown, highly advanced civilization. This civilization is one to which every human being on Earth has a right to have knowledge of and make peaceful contact with.

My personal opinion is that the authorities can either join and assist the public in making peaceful, open contact with the mysterious people operating the UFOs and/or flying saucers or get the hell out of the way!

W.D. Clendenon

viii

Book 1
TABLE OF CONTENTS

CHAPTER ONE: In the Beginning

During the latter part of World War II while serving in the United States Navy as an Aircraft Recognition Instructor in the Southwest Pacific area of New Guinea and the Philippine Islands, I had on different occasions heard vague reports or rumors of so called foo fighters or ghost rockets as UFOs were referred to throughout the world in those days.

I was interested not only because it was my official duty as a Navy Aircraft Identification Instructor to keep up with all of the latest information on aircraft that either the Japanese or allied military forces introduced into the combat areas, but because the UFO reports were coming from all parties in conflict from around the world. This, therefore, presented a real life mystery that continues to be a mystery to our world even to this day, forty years later.

After I was discharged from the United States Navy at the end of World War II, I continued to be interested in aviation and aerospace subjects in general. My interest in general aviation and aerospace subjects resulted in my continuing to build model aircraft as I had when I was a boy. I also took private flying lessons to become an amateur pilot. Then, too, I had an on-again, off-again membership in the Experimental Aircraft Association. These things continued to develop my interest in all types of flying machines no matter how successful or unsuccessful the design or its origins. In other words, my mind was open to all new ideas no matter what the source if it pertained to aircraft of any sort, I was particularly interested in any aircraft that showed promise of vertical takeoff and landing flight and even flying submarines, and so I researched odd types of designs, both past and present.

Toward the end of 1946, while staying at my grandmother's house in Alabama, my aunt who lived down the hill began to tell members of the family that she was seeing strange lights go over her house every evening. She described the lights as ball-shaped and somewhat greenish.

But no one would pay any attention to her. The first time I heard my aunt describe what she was seeing, my ears perked up because of the reports of foo fighters that I had heard previously. I told her to tell me when she was going to be home the next evening and I would come down with her so we could both watch and see if the lights came over her house again.

The next evening after I had spoken to my aunt about the strange greenish balls of light flying over her house, I started down to her house to sit with her and watch. It was just before dark and as I looked up, I saw my first UFO. It was a greenish ball of light traveling horizontally at high speed and heading west. It went directly over my aunt's house. As I looked down, I saw my aunt coming up the hill, shouting and waving her arms, pointing into the air. I acknowledged that I had seen the object and, from that time, I was hooked on the UFO subject in earnest.

In the next few years, I was to have many more sightings of these strange greenish lights in the sky. The green was the shade of the flame of an acetylene torch, a coppery green. There was no sound during any of these sightings. There were, at times, parallel streaks of what appeared to be vapor trailing behind the balls of light. The vapor appeared in spurts as the UFO's speed decreased. As the UFO speeded up in its forward flight, the ball of light cut out, as though a switch had been flipped to the off position.

It is important to understand that these unidentified flying objects that I was observing in the balls of light were travelling horizontally. That is, they were not moving in an arc similar to a meteorite. Not only that, these unidentified balls of light were at a very low altitude. Of course, I could not be exact as to the correct altitude, although I would estimate 3,000-9,000 feet, as I had no way to gauge it.

Since I travelled a good deal in the building trades, my work took me to Lima, Ohio where I was employed in an oil refinery as a boilermaker, repairing and building different types of structures such as oil storage tanks, bubble towers, and so on. This refinery was adjacent to an army depot where tanks, armored cars, and other military equipment was stored.

Just before Thanksgiving, 1949, on a bright, sunny day, I was working with a crew atop a bubble tower. Our foreman had gathered us in a group and was explaining the things he wanted done in regard to putting the trays in the bubble tower. While he was talking, I looked up and saw a UFO hovering in the air. To this day, this UFO was the largest I have ever seen and the first daylight sighting for me.

The object was disc-shaped and curved across the top. It appeared to be about 100 feet across, estimated by one of the engineers who made rough calculations. However, when I first noticed it, I watched it and thought to myself, "If I say anything, the darn thing will disappear and they'll all think I'm nuts."

So, I merely said, "Fellows, look up and tell me if you see anything."

They did and I watched their faces instead of looking at the ship. They all stood there with blank expressions on their faces and some of them with their mouths open. Finally, when I was fully convinced they were seeing what I saw, I turned my attention back to the UFO. It appeared to be dark in color. It was not close enough to distinguish but you could tell it was curved across the top and disc-shaped.

We watched it for several minutes. I took time to note that there were other people in the refinery also looking at it; some cars had even stopped on the highway. Suddenly, a white vapor came out of the craft, with a space between the vapor and the edge of the ship. The vapor came out not like contrails from aircraft, but in a cloud of steam-like vapor. As soon as the vapor began to disappear, the craft moved off slowly at first. It gathered speed and it just seemed to evaporate. There was no sound.

We all discussed the UFO sighting at some length and one of the fellows said it must be some kind of jet. I explained to him that it couldn't have been a jet as we know them because we didn't have anything like that which could hover motionless in the sky without making any sound.

When I got home from work that day, I turned on the radio and listened for a report. Other people had seen the same UFO that we had.

After that there was no possibility of abandoning the quest for truth about UFO's.

NOTE: Danish UFO forming cloud-like vapor; a similar propulsion side-effect was witnessed during UFO sighting described on page 3.

CHAPTER TWO: UFOs and Adamski

In 1951, I was employed as a boilermaker at an Air Force wind tunnel in Tennessee and worked on and off there for nine years. While working there, I talked with many engineers and professional people in the aerospace field and learned about aerospace subjects. I read many of their technical magazines. I was employed in additional construction on the wind tunnel as well as the installation, construction, and repair of turbines, boilers, heat exchangers, condensers, plenum chambers, gas ducting, and storage tanks for liquids. Little by little, I was gaining the basic knowledge which I would need later in researching the propulsion system for an Adamski-type UFO Scout Ship.

While working at this Air Force facility, I engaged in the hobby of spelunking, cave exploring. I joined the National Speleological Society. The small group of cave explorers to which I belonged came from a wide variety of backgrounds, such as construction workers, businessmen, doctors, mathematicians, and physicists. Since these people were either employed with the Air Force facility or nearby, we spent much time together as a group exploring the caves on weekends and at other times.

One time we met to explore a cave between the towns of Tullahoma and Shelbyville, Tennessee. After spending the entire day in the cave, we came out at dusk.

As we exited the cave, we faced the setting sun. Although the sun was down behind the tree line, it was still very light. As we walked back toward our cars, I glanced up to my right and saw a brilliant ball of light hovering above the trees about a quarter mile away. It was very large. I noticed the others had seen it too and decided to wait to see what their reactions were going to be.

The physicist said, "Look at that bright star!"

"That's no star," I told him. "It's too low and too damn close to be a star. It's a UFO."

Nobody said anything. I no sooner stopped talking,

when the ball of light dropped straight down and, as it did so, it changed colors and intensity of light, and disappeared behind the trees.

As we stood there trying to decide just what it was, the ball of light came straight up from behind the trees and hovered again. There was no mistaking this time that it was a UFO; one of the best sightings I had ever had. I was very happy because I had seven witnesses with me, including two scientists. However, I was to be the only one who would stand his ground in regard to this sighting.

As we stood there, I decided to get closer and took off across the cow pasture toward the object. As I got closer, the light would let me get within a certain distance of it, then move away, still above the trees to our right.

I walked along the tree line, keeping the UFO in sight and by this time, it was getting dark. I got to my vehicle and followed the UFO as best I could along the dirt road. The fence was in the way and I couldn't find an opening in it. I came to a curve in the road and still had the UFO in sight. I parked the truck and got out and followed the UFO on foot through the wooded area. Again, the UFO would let me get just so close and then it would move away. I noticed that the UFO could maneuver in a most extraordinary fashion and that the lights would change in intensity and color almost constantly.

The absence of any sound was outstanding especially when I realized the size was from 50-75 feet in diameter, just the ball of light, not the object in the center of the light. Although I could make out a form, there were no discernible details. The form within the ball of light was smaller.

When I rejoined my companions, I asked the others about reporting it. Immediately, they told me they would not get involved in the publicity as they had professional reputations at stake. I went to the sheriff's office and made a report. However, it was fruitless. In the next few days, I made a trip to the Air Force base at Smyrna, Tennessee. I went to the base and made a report to the person in charge but again, it proved fruitless.

In the meantime, I continued to be interested in building

6

model airplanes. I specialized in circular air frames and air foils and I did considerable research and wrote to people in the EAA, talked to many engineers and scientists at the Air Force facility, and put all my findings together. I showed it to one of the engineers at the Air Force facility and was informed that they had a project much like it. Later, I discovered it was the AVRO Project that we were to hear so much about in the future.

In reference to the famous UFO Scout Ship photo taken by George Adamski in December, 1952, I noticed the UFO photograph and was immediately interested in it. There was no information with it other than the name of the person who took it. As I examined it, I wondered if this wasn't a real UFO, what had the man used as a model.

At first I thought it looked much like a hi-fi speaker, or at least, parts of a speaker, so I set it aside and gave it no more thought at that time.

Then, one day I saw a television program featuring the man who took the photograph, George Adamski. He was telling his story about having met a man who had stepped out of a UFO like the one in the photograph. Mr. Adamski stated the man claimed he was from Venus. Although I was skeptical, the picture still haunted me and I couldn't forget it. When I discovered that George Adamski had published a book in collaboration with Desmond Leslie called, *Flying Saucers Have Landed,* I wrote to George Adamski and purchased a copy of the book.

When the book arrived, I read it several times. Some of the information about the Mercury Propulsion Systems of ancient India was familiar to me as I had seen it in other periodicals. This book, however, was more detailed. But, when I got to Adamski's section of the book, I had some reservations accepting his story as he told of meeting a man from the UFO in the California desert and of talking with him. Although I didn't disbelieve the whole book at this time, I just let it cook in my head for some time while I continued building model airplanes, circular in form, and experimenting with them.

Finally, I decided to examine the photograph again,

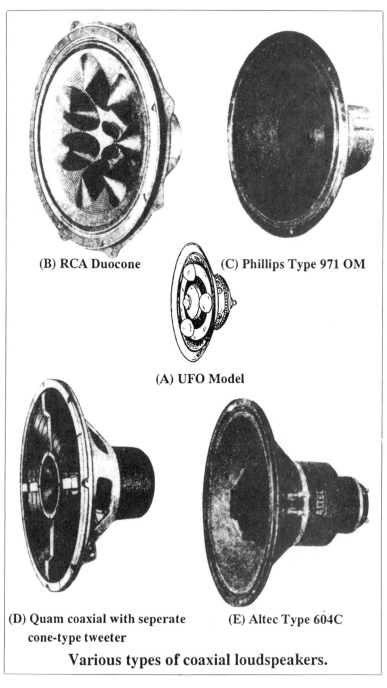

(B) RCA Duocone   (C) Phillips Type 971 OM

(A) UFO Model

(D) Quam coaxial with seperate   (E) Altec Type 604C
    cone-type tweeter

Various types of coaxial loudspeakers.

If form determines function the resemblance of the Adamski UFO Scout Ship to a Hi-Fi speaker may not be a coincidence.

Clendenon holding model UFO and research drawings
in August, 1959.

9

simply as a hobby. I had the photograph enlarged so I could use it in my research. However, I soon discovered that I needed a clearer photograph because the one in *Flying Saucers Have Landed* was not clear. I wrote to George Adamski and purchased a copy from the original negative.

Over the years I used the UFO photograph in my research. I counted the UFO's parts. I noted their form and arrangement. I noted the shading and the color of the craft. I proceeded to experiment with models shaped like the one in the photograph. I was determined to discover whether or not this was a real UFO, a hi-fi speaker, sun helmet, etc.

Although my mind would not let go of the UFO photograph, my good sense refused Mr. Adamski's section of *Flying Saucers Have Landed* as published. As I read and reread the book, studied the UFO photograph, and experimented with models, my mind formed a master plan for researching the photo.

That plan was this: Keeping an open mind, I would research the ancient technical information (plus UFO photo) in *Flying Saucers Have Landed* and would then apply it to known advanced aerospace craft technology for comparison.

I felt that correctly identifying the object in the Adamski UFO Scout Ship photograph was the key to unlocking the UFO Mystery.

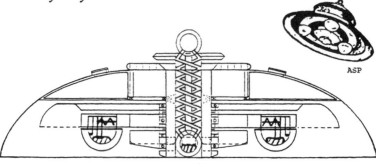

ASP

FIG-1.
SIDE-PLAN.

Basic model aircraft (VIMANA) gas turbine powered (electromagnetic)

CHAPTER THREE: UFOs Overhead

I made a number of sketches, mountains of notes, sat around, and thought about all this for awhile before I wrote Adamski a letter and told him honestly that I had trouble accepting his story. I did accept his UFO Scout Ship photograph as being real (up to a point) because of the experiments I made with the aircraft models, the research I did on the Scout Ship photograph, and from studying Leslie's notes on the ancient Sanskrit writings about the Mercury Propulsion systems. So I told Mr. Adamski that I would keep an open mind. I told him I thought it only fair that he should have a copy of my efforts to decode the information in *Flying Saucers Have Landed*, so I sent him the material.

I waited for two weeks or more wondering if Adamski was going to answer my letter. At that time, I lived in a little house trailer and about 9:00 p.m., I went outside. It was in the summer and I took a cup of coffee with me. The night was starry and I was studying the stars when, suddenly, I saw a greenish ball of light, the same type and size that I had seen years before in Alabama. The ball of light streaked back and forth over my house. The yard was surrounded by trees with a space in the middle. All night long the UFO travelled back and forth, from different directions directly above the yard. I watched this activity until dawn, realizing that the overflights formed a triangle.

Later in the day, I looked up some information on navigation which stated that an aircraft pilot sometimes used *triangulation* to fix a point or to establish his correct position. I concentrated on information on triangles and myths, pyramid shapes and triangles, or hidden meanings. I found information to the effect that the form of a equilateral pyramid or triangle meant completeness. I deducted two things: I had a form that, according to myths meant *completeness,* and aircraft pilots used triangulation to fix a point in regard to finding a position. Apparently this being piloting this aircraft was fixing my house as his point of reference and they wanted me to *complete* what I had started.

The next night I stayed outside all night, watching and again the same thing happened. The green ball of light came over my yard and went through all of those triangular maneuvers and this time, my daughter came outside with me. The two of us, along with the family dog, stayed out there, watching, for a couple of hours, as the ball of light went back and forth, first one way and then the other. Finally, I stepped inside the house to get another cup of coffee.

As I started back outside, I saw a flash of light and my daughter began yelling, "Daddy, Daddy, Daddy." The dog was barking as I hurried back outside. The dog and my child were facing a small ball of light, about three feet in diameter. It had a wobble to it much like a gyroscope and the light was yellowish-orange. However, it fluctuated between this color and white. I was about 15 feet from the ball of light and my daughter was 5-6 feet from it. The dog went much closer. At first, I had the strong impulse to grab it, so I stepped closer to it. As I did so, the dog snapped at it and I saw that all the dog's hair was standing straight up. I am not sure whether it was from fright or an electrical field around the light. As I stepped closer, the object shot straight up and disappeared. Had I blinked, I would have missed it.

There was no sound, nor any heat. Neither my daughter nor I were frightened, just very excited. We went into the house and I had my daughter draw what she had seen. (I studied that sketch many, many times and found out later on that it was quite useful, as simple as it was.) After that, we went back outside. The UFO triangulation maneuvers continued for awhile and then abruptly stopped. However, the activity continued for a few more nights.

The ball of light was about one to one and one-half feet above the ground. It made no marks or disturbances on the surface of the ground. I looked for burn spots in the grass or anything, but nothing existed. I made notes about the incident and kept my daughter's sketch.

On the fourth night, during one of the overhead passes, the UFO descended in a long shallow arc and disappeared. These maneuvers were different as the oval-like light changed size, shape, and color. The UFO apparently landed.

I did not go to look for it because it would have been impossible to judge where it had landed.

About an hour after the UFO sighting, I was still outside in the yard hoping to see more UFOs. However, the UFO activity had come to a halt.

Shortly, a car drove past my home slowly as though the driver were looking for something or someone. The second time the car passed by, I waved. The driver responded but kept going, returned, and stopped. I walked over to the car and saw that it contained two young people, a man and a woman, Caucasian, in their mid-twenties. The three of us exchanged greetings and the man asked directions to the main highway. After I gave them, the young woman spoke up and commented on the starlit night sky and how beautiful it was. I replied that I had been observing the night sky for the past few nights and had seen very interesting sights. At this remark, the young man spoke up and stated something to the effect that with enough time and patience one may see some wonderful sights up there.

I agreed and the young couple thanked me for the help and drove off.

**Pulsating Ball of Light Seen on
an Evening in August, 1958
Hovering approximately 1 foot
above ground**

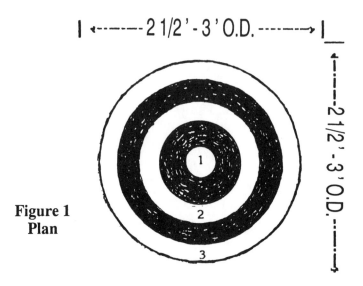

**Figure 1
Plan**

Key: **1 - small bright white ball of light
2 - bright white halo light
3 - bright white halo light**

NOTE: I was informed by Adamski that the above
Pulsating Ball of Light was a little registering disk
remotely controlled for observation work. Our military
forces also employ R.P.V.'s or Remotely Piloted Vehicles
for much the same reasons.

CHAPTER FOUR: The Visitor

Eventually I moved from Tennessee back to the state of Washington and lived in Seattle. George Adamski published a third book, *Flying Saucers, Farewell.* Of course, I bought it and read it, studied it for a while, using it the same way I did his other two books. I mailed my latest drawings and notes to Mr. Adamski and this time I received a written reply from him in just a matter of days. He wished me well in my research, encouraged me not to give up, and I truly thought that this would be the end of our correspondence.

While I was keeping an open mind on the case of George Adamski and his UFO photographs, I still had not accepted all his claims 100%. There was no doubt in my mind that he had made contact with the beings flying these mysterious ships which came over my house. But, in the back of my mind, there were doubts in other areas.

The heavy evangelistic philosophy of Adamski's California group bothered me. Although I didn't mention my doubts to Adamski at this time. However, my research findings in regard to Adamski's UFO Scout Ship photograph left no doubt in my mind that there was something in the Adamski UFO case that warranted further investigation.

Aug 28, 1961

Dear Mr. Clendenon Jr.:

Have received your most welcome letter and your roll of drawings and I wish you all the luck in the world for the hard work you have done; I have the first drawing put away somewhere as I sold out at Palomar Terraces, so everything is packed as I might be making a move to another state or probably Mexico in the next thirty or sixty days. You will always be able to keep in contact me through this phamplet. I might even have more information on the propulsion business when I once settle down at my permanent address.

Sincerely yours,

15

A few days after I received the preceding letter from Mr. Adamski, I was home alone. When the doorbell rang at 9:00 p.m., I went to the door. I saw through the full length glass window of the outer door that a man was there. The man, 5 ft. 6 in., well-dressed and wearing a tan topcoat, looked to be about 30 years old.

When I asked him what he wanted, he told me that he was taking a political survey of the people in the area to see whether they voted on the Republican ticket. I told him I voted independently, depending on the individual candidates, etc. The man nodded his head and I shut the door, thinking that was that. However, I experienced a very strong impulse. Something in my head said, "Ask the man inside now."

Confused, I did so. I sat down on the couch and my visitor sat on the easy chair. I stated that since he was involved in public relations, perhaps he could advise me on how to generate interest in an invention I had. When he asked what the device was, I told him it was an aircraft that could take off and land vertically. At this point, he asked me if I was having any technical problems with my invention. I told him I was but I thought I could work things out in time. We discussed some of the problems with the invention and soon I realized that this man knew much about my invention.

He stated that I should try to interest the government and aerospace firms, such as Boeing Aircraft, even if I had to go to Bill Boeing himself. I told him I would try. (I did, too, but got nowhere.)

We talked for almost an hour and he was interested in the type of books I had laying about regarding aviation, science, UFOs, etc.

At no time did either of us mention George Adamski, UFOs, or anything like that. We talked about my invention. As he stood up to go, he turned and began jotting some notes on index cards he had withdrawn from his pocket. I noticed that he wore dark-framed glasses although I instinctly knew he didn't need them. However, the thing that impressed me the most was his face. His skin was smooth

16

Likeness of Alien visitor, Sept. 1961

as though he never shaved in his life. His skin reminded me of a baby's skin. When he smiled, his teeth were perfect and very white. The color of his skin was brown, like an Indian; his hair was dark and trimmed in a business-like manner. He looked almost too perfect and it bothered me.

The stranger left; I did not try to follow him and I never saw him again.

However, I felt keyed up the way things were beginning to shape up. But, not in my wildest dreams did I imagine what events would transpire in the next few days.

Once more, after the man had visited me, I felt the urge to walk out in my backyard. It was nearly dusk but still light enough to see well. While standing in the backyard, facing north, I saw a brilliant white ball of light. It wobbled as if out of control, but straightened itself in a westward line. I saw that it was accompanied by short spurts of white vapor. The ball of light made a right angle turn due south, and proceeded toward Bremington, when the UFO reached a position parallel with my position. It made another right angle turn and came eastward toward me. As it came closer to me, the white light turned to rust-orange and was pulsating on and off.

The details of the UFO's structure now became quite clear and looked identical in every way with the Scout Ship photographed by Adamski in California in December, 1952.

I had just been shown that Adamski's UFO Scout Ship photograph was real.

As the white Scout Ship came closer to where I was standing, I could make out the details of the 35'-40' outside diameter of the ship with ease. The ship had a rust orange light at the tip of the ball atop the cabin. Around the top edge of the cabin, there was a rust orange halo and another orange halo around the extreme lower flange of the UFO. Another rust orange light was on the bottom side in the very center. All these lights were pulsating on and off as one, not independently.

There was no sound but while standing directly under the UFO with it about 100-150 feet above me, I could

18

Name __CLENDENON, William Duey, Jr.__
(Name in Full, Surname to the Left)

__273 18 62__ Rate __S1c, V-6 USNR.__
(Service No.)

Date Reported Aboard: _____28 November 1944._____
FLEET TRAINING COMMAND SEVENTH FLEET
FLEET POST OFFICE
(Present Ship or Station)
SAN FRANCISCO CALIF,
AATC, Navy 134
(Ship or Station Received From)

6 Jan. 1945: Transferred this date to
Anti-Aircraft Training
Center, Navy 3205.

Auth: Com7thFlt Ltr. Ser. 15932,
of 31 Aug. 1944.

ESPECIALLY TRAINED AND QUALIFIED AS
INSTRUCTOR IN AIR AND SURFACE RECOGNITION.
HAS PERFORMED OUTSTANDING SERVICE IN THAT
CAPACITY.

A. G. W. McFADDEN
COMMANDER, USN

Date Transferred __6 Jan. 1945__

To __Anti-Aircraft Training Center,__
Navy 3205

A. G. W. McFADDEN, Comdr., USN
Signature and Rank of Officer Authorized to Sign

Date Received Aboard: __20 January 1945.__

__A.A. Training Center, Navy 3205.__
(New Ship or Station)

__Fleet Training Command, 7th Flt.__
(Last Ship or Station)
C. T. Budgman LT(jg) U.S.N.R.
H. S. PRICE Lieut. USN?R.
Signature and Rank of Officer Authorized to Sign

ORIGINAL
FOR SERVICE RECORD

Certification of Qualification for Instructor in Air and Surface Recognition for William D. Clendenon, Jr.

Fig. 1.

←————— 35' - 40' - O.D. —————→

Fig. 1

UFO as witnessed by Clendenon at dusk, September, 1961. Altitude approximately 100 to 150 feet from ground surface.

KEY:

(1) Orange ball of light
(2) Orange cabin halo
(3) Orange outer flange halo
(4) Orange center bottom light

NOTE: The above orange electrical lights and halos pulsated in unison. No sounds or noises originating from the aircraft were detected although I sensed some heat.

detect some heat but no odor. The little ship now moved so slowly above my head that I could barely detect any forward movement. I could, however, see the turbine wheels revolving and some hatch-like outlines on the underside of the ship in the center area. This UFO sighting was the best in every way. I learned much in a short time. Portholes or windows were very easy to see. The little Scout Ship picked up speed and moved eastward from my backyard toward Sandpoint Naval Air Station and I felt the personnel at the Navy Air Base just had to see it, but they told me the next day when I checked with them that they hadn't seen anything relating to a UFO during that time. However, many people in my area reported the UFO sighting to a radio talk show the evening of the sighting. Those people received little positive response. The radio show's host made remarks like, "We have another weirdo in the sky, folks!" and so it went.

However, I knew that Adamski had indeed been involved with the UFO Scout Ship. Of that fact, I no longer had any doubts!

Adamski U.F.O. photo 1952

CHAPTER FIVE: Adamski's Death

After the encounter and sighting of the Adamski-type UFO Scout Ship in Washington state, I was to have several less dramatic UFO sightings when living in Oregon. These were in the form of pulsating balls of light.

While in Oregon I attended a UFO lecture given by a person who has and still is doing damage to ufology. This individual claims to have had contact with an extraterrestrial being at the Pentagon. He also presented in public a specimen of marine life and labelled it an *alien form of life.* (This was a specimen of sting ray. In that sense, it was an *alien* form of life as compared to *humans.*)

This conduct is only one of the many reasons why the American public should demand immediate open Congressional hearings on UFOs. Anyone who lies to Congress during open hearings is subject to criminal prosecution. I would welcome open hearings on UFOs as I have stated and written many times in the last 40 years. Congress keeps evading the UFO issue; only pressure from the American people will bring results. I am willing to testify any time Congress holds hearings about UFO's.

Open hearings would clear up the case of George Adamski. The confusion about the notorious Straith Letter, which was supposedly sent to George Adamski by R.E. Straith of the Department of State, would be clarified.

The matter of this letter is covered by English UFO researcher Timothy Good in *George Adamski, The Untold Story* and *Above Top Secret.* It seems odd to me that these two fine reference books so full of facts met with so many obstacles with distribution throughout the world.

In 1964, events happened which confused me in my research. *Mis-information* by the United States government, some of Adamski's co-workers, and perhaps even the crews of the UFO Scout Ships seem to have set a cover-up in motion which continues to the present time.

In 1964, while living in Portland, Oregon, Mr. Harold C. Moody of Milwaukie, Oregon, who was the project personnel manager for the AVRO-VZ-9V AVROCAR project, heard about my UFO propulsion research and came to see me.

The VZ-9V was a disc-shaped aircraft that could not rise vertically more than a few feet off the ground after which vectored thrust gave it forward motion. However, the VZ-9V project was canceled, so we are told, because of stability problems in forward flight.

Mr. Moody was interested in my ideas of rectifying this stability problem. I told him I would employ the gyroscopic-like instability inherent in the disc turbine design and control it by a computer because of lag time and control. Mr. Moody was disappointed that the AVRO project was abandoned; he knew some people who might be interested in my idea.

Up until 1964, Adamski and the strangers flying the UFO Scout Ship encouraged me in my efforts to duplicate the Scout Ship propulsion system in model form.

After the UFO sightings and my personal visits from the strangers, I realized that my personal relationships were deteriorating. No one close to me understood what was happening, the situation I was experiencing, nor the rationale for putting more and more money into the UFO research without any apparent return.

I stood alone in my UFO research efforts. The general consensus of those around me was: Who was I, a comparative nobody, to challenge the government and professional people in the Aerospace field? And what did I know about UFOs as compared to them?

However, I suffered the most for having the nerve to state in public that I had seen a UFO Scout Ship; the one I had seen being exactly the same as the one shown in Adamski's photograph taken in the 1950s.

My next problem was to raise funds to patent and copyright my work. I had to find a way to inform the public about the George Adamski case but still not offer too much of my research information so it could be stolen by someone.

I thought that publishing an article might raise funds and

The United States and Canada once tried to build a flying saucer. This picture of the disc-shaped Avro car, developed by AVRO Aircraft Ltd. of Canada, was first released in 1960. The project was abandoned in 1965 because the craft could not rise more than a few feet in the air with stability.

contacted Ray Palmer of Palmer Publications, Amherst, Wisconsin. I explained what I was doing and discussed the problem of protecting my research. Ray Palmer and I decided that I would send him some of my notes and drawings and he would publish it in a way that it would draw the public's attention while maintaining some protection for my work. He scrambled some of the notes, published them with a few drawings; the results were satisfactory. It drew attention to my work, but did not give away too much.

In essence, Ray published worthless text with proper drawings for the propulsion system. When I first saw the publication results, I was shocked. I thought my credibility was ruined. However, Ray Palmer knew what he was doing. I received more mail in the next 18 months than I ever expected. My biggest surprise was a letter from George Adamski.

On September 10, 1964, Adamski left home for a lecture tour in eastern United States. He spent some time in Washington, D.C. to meet with some people from the Space Administration. While there, he stayed with Mr. and Mrs. N.E. Rodeffer, Silver Springs, Maryland. Meanwhile, I had sent Adamski additional pieces of my research, including drawings. However, my material arrived at his home after he had left for the lecture tour. A close associate of Adamski, Mrs. Alice K. Wells, informed me that she would hold my papers for him until he returned home. A few days later, Mrs. Wells wrote me the following letter:

September 23, 1964

Dear Mr. Clendenon:
                I talked to Mr. Adamski on the phone last night and he said to send your material on to him and he will see what he can do with it in Washington D.C. If you would like to get in touch with him while he is there you could write to him in care of, N.E. Rodefer, 12905 Falmouth Dr., Silver Springs, Maryland.

Sincerely

Alice K. Wells

Alice K. Wells

25

Up until this time, nobody had indicated to me that I should not make my findings public. Adamski and his associates knew and understood about the 1964 article in *Flying Saucers*. When Adamski came to the phone, he indicated that he had been informed that though there were some errors in my Scout Ship Propulsion research, he had been told I was correct in my basic findings. He also indicated that the situation had changed and there were problems in the UFO program. I told him that Ray Palmer had suggested my research be published in a book. Adamski told me to suspend my research efforts for awhile and to forget about the book. I was very confused. I could not understand the sudden change in attitude. When I asked him about this confusing change of attitude, he told me that we would talk later. He told me that we would both be in serious trouble if I didn't cooperate. I agreed to abide by his wishes but the whole thing bothered me deeply. It still bothers me to this day for it is an unresolved issue.

Up to this point, my dealings with Adamski had been centered around the Scout Ship Propulsion research and that was the bulk of our discussions. I trusted him because he had proven to me that he had contact with the pilots of the UFO Scout Ship. He knew about my visits from the aliens before I mentioned them to him. He was a credible man.

After Adamski returned from his tour, he wrote to me on October 31, 1964, asking me if I would send him everything I had on my propulsion research as he knew someone in Mexico who wanted to see my work. He also mentioned that he was going to visit some Mayan Ruins in Yucatan where he expected to find something about which the UFO Scout Ship pilots had told him.

After he returned from Mexico, he wrote to me on February 16, 1965 and told me that the trip was a success but that someone in Mexico had told him about an article in FATE magazine about me, that there would be more articles of the same kind, and that I was involved in N.I.C.A.P. To this day, I am still confused about these statements for these reasons:

1. I do not know anyone in Mexico.

2. I have no knowledge of any articles either by or about me in FATE magazine.

3. I have never joined a civilian UFO organization.

Adamski further told me that he knew a man in South Africa that had published his UFO research findings, and because of it nobody knows the fate of the man. He just disappeared.

During his last trip east while staying at the Rodeffer's home, I spoke to George Adamski on the phone, trying to clear up the confusion of his letter of February 16, 1965. The man was troubled; I could tell by the tone of his voice. Again, he told me to watch out for trouble.

The next information I received about George Adamski was that he had passed away in Tacoma Park, Maryland, on April 23, 1965.

I received a letter dated May 10, 1965 from Mrs. Alice K. Wells informing me of Adamski's death on April 23, 1965. She also stated that she would be coordinator for the George Adamski Foundation and that I should deal with her in matters pertaining to my UFO research dealing with the case of George Adamski.

Mrs. Wells could just as well said, ''Let the games begin'' for with Adamski's death, games began in earnest and Mrs. Alice K. Wells was a strong player on the UFO field.

Some people, and I was one of them, wondered if there was any mysterious reason for Adamski's remains to be interred at the Arlington National Cemetery. Was there some information concerning a possible outstanding contribution to our nation that was not well-known? With this in mind, I contacted the office of the Superintendent of the Arlington National Cemetery in Arlington, Virginia.

The letters I received made me wonder if there was a mystery regarding Adamski's death, especially the sudden change which seemed to settle over the Adamski case after his September, 1964, visit to Washington, D.C.

The following letters illustrate my point:

J. W. FULBRIGHT, ARK., CHAIRMAN

JOHN SPARKMAN, ALA.
MIKE MANSFIELD, MONT.
WAYNE MORSE, OREG.
RUSSELL B. LONG, LA.
ALBERT GORE, TENN.
FRANK J. LAUSCHE, OHIO
FRANK CHURCH, IDAHO
STUART SYMINGTON, MO.
THOMAS J. DODD, CONN.
JOSEPH S. CLARK, PA.
CLAIBORNE PELL, R.I.
EUGENE J. MCCARTHY, MINN.

BOURKE B. HICKENLOOPER, IOWA
GEORGE D. AIKEN, VT.
FRANK CARLSON, KANS.
JOHN J. WILLIAMS, DEL.
KARL E. MUNDT, S. DAK.
CLIFFORD P. CASE, N.J.

CARL MARCY, CHIEF OF STAFF
DARRELL ST. CLAIRE, CLERK

United States Senate
COMMITTEE ON FOREIGN RELATIONS

June 7, 1965

Mr. William D. Clendenon, Jr.
P. O. Box 926
Portland, Oregon 97207

Dear Mr. Clendenon:

I appreciated your very kind letter of June 1 and thank you for giving me the benefit of your views on several contemporary issues.

I have inquired concerning the burial of George Adamski. A man by the name of George Adamski was buried in Arlington National Cemetary on April 29, 1965, in Section 43, Grave 295. He had served as a private (#349-778) in the U. S. Army in 1918-1919. The information concerning his possible outstanding contributions to our Nation is not available. However, according to law any man who has served honorably in the Armed Forces of the United States may be buried in a National Cemetary.

For further details I suggest that you contact his next of kin -- his niece Alice K. Wells, 314 Lado de Loma Drive, Vista, California -- or else Funeral Director W. Pumphrey, Silver Springs, Maryland. Mr. Adamski passed away in Tacoma Park, Maryland, on April 23.

I appreciated hearing from you and send best regards.

Sincerely,

Wayne Morse

Wayne Morse

WM:sdm

28

OFFICE OF THE SUPERINTENDENT
ARLINGTON NATIONAL CEMETERY
ARLINGTON 11, VIRGINIA

18 June 1965

Mr. W. D. Clendenon, Jr.
P. O. Box 926
Portland, Oregon 97207

Dear Mr. Clendenon:

Please be advised that the records of this office
and the master file of the national cemetery system
indicate that George Adamski is not interred in
Arlington or any other national cemetery.

Sincerely,

J. METZLER
Superintendent

29

30 June 1965

Honorable Wayne Morse
United States Senate
Washington, D. C.

Dear Senator Morse:

This is in reply to your inquiry of 29 June
regarding the interment of the late George Adamski
in Arlington National Cemetery.

It is deeply regretted that the burial record
of George Adamski was misfiled in this office
under George Ādanski. I can only offer my sincerest
apologies for this error and the inconvenience it
may have caused.

The records of this office indicate the
following:

Private George Adamski, U.S. Army; serial number
3649778; enlisted 24 July 1918; discharged 15 January
1919; died 23 April 1965 in Takoma Park, Maryland
and was interred on 29 April 1965 in Section 43
Grave 295.

Mr. Adamski was eligible for burial in Arlington
under the provisions of Public Law 526, 80th Congress,
(62 Stat. 234) as amended by Public Law 86-260, 86th
Congress, (73 Stat. 547) as indicated in paragraph 1
of the enclosed brochure.

Sincerely,

JOHN C. METZLER
Superintendent

The actions of Mrs. Wells and others close to Adamski reinforced my thoughts. Until the end of 1965, after Adamski's death, I received an average of one letter per month from Mrs. Wells plus the Cosmic Bulletin published by the George Adamski Foundation. Of course, there were phone calls going in both directions.

I noticed a trend, a pattern was forming in most of Mrs. Wells' correspondence to me. It began to be heavily laden with philosophical and religious terms such as "brotherly love" in addition to quotes from the Bible. There were also hints and remarks pertaining to technical information such as "nuclear power is being used in UFO propulsion." She also made frequent remarks that the U.S. government already had working models of the UFO Scout Ship indicating that she thought I should stop my propulsion research. However, contradicting this attitude, she also wished me good luck in my research efforts in every closing of her letters.

The results were that I was gradually being led astray on one technical tangent after another. Confusion was settling in and Mrs. Wells' conduct regarding this misinformation program angered and frustrated me. There was something else I noticed in some of the pamphlets being sent out by Adamski's followers. In addition to the genuine photographs, there were obvious fakes intermingled with them. Little by little more fakes appeared and less genuine photographs. When I pointed this out to Mrs. Wells by phone and letter, the result was that the apparent deliberate confusion campaign for my personal benefit was accelerated.

One afternoon right after work while I was enjoying The Three Stooges on television while home alone, a very distinct male voice cut through the T.V. program without distortion of any kind. The voice said, "Bill, keep your mouth shut." I was not frightened but it sure grabbed my attention.

A few days later, I went to the local television station to find out if they had any information. I asked them if it would be possible to project a voice through a selected tele-

31

vision. I was told, to my surprise, that it would be possible but that the person would have to have some "highly sophisticated electronic equipment" with which to work. The engineer I talked to indicated that the U.S. government would be his first suspect.

But, I had other suspects in mind. The strange people flying the Adamski-type UFO Scout Ship!

I received this letter from Mrs. Wells:

October 19, 1965

Dear Bill Clendenon:

You have been patient this long - so do not become impatient now and spoil everything you have worked for. You are in a position to either help yourself or hurt yourself. Why not wait a while longer and see what developes ? By all means I would ignore the offer from the magazine. When Van den Berg told his story about his motor to an African publication and it was republished in England - that was his finish - no one seems to know what happened to him. At least you now have your freedom and your family - so be patient.

There is no magic word, or presto for success, everything worth while has its time for maturity. No one tells the space visitors what they should do - that would be utter foolishness for they have wisdom far beyound ours. You are developing your own strenght of character and that is important if we are to live a full happy life. The old expression ' Discresion is the better part of valor,' is just as true today as it ever was.

No one seems to know where Rick Williamson is or what he is doing. He violated certain trusts that were place in him by the Brothers and George Adamski, so he has a price to pay. Oblivion perhaps !!

I trust the tone of my letter is not frightening but I have had many experiences with Mr. Adamski and if you can benefit from them, I am glad to share them with you.

Thank you for keeping me informed and I wish you every success.

Sincerely

Alice K. Wells -

Alice K. Wells

CHAPTER SIX: Confusion

I tried to get the subject of UFOs out of my mind completely for it was a big, expensive, time-consuming project which seemed to draw trouble from all sides. However, something deep inside me kept me fascinated with UFOs in general and the Adamski case in particular. So my research efforts continued.

I made another trip to California, went to Adamski's home, and paid a short visit to Mrs. Wells, one of many contacts with her. During this visit, she showed me some UFO photos which I had not seen before and some other UFO artifacts. A large, beautiful, colored picture of a UFO Scout Ship hung on the wall in the living room, two plastic Scout Ship models, very authentic and detailed, hung on either side of the dining room door, and a lifesize color painting of the man Adamski met in his first contact in the California desert in 1952 hung near that.

During our guarded conversation, she told me to keep her informed of my location. When I asked her if I could buy one of the models, she refused. She gave me some UFO photographs and we discussed them. Mrs. Wells had written to me previously about Rex Heflin's UFO photographs taken near Santa Anna, California in 1965. She told me that this particular UFO was a Saturn-type Scout Ship. I had noticed the likeness of Heflin's UFO photos to the sideview drawing on Diagram 9 of Adamski's book, *Inside the Space Ships*. I pointed out that the smoke ring UFO photograph, which was one of four UFO photographs Heflin took of the Saturn UFO, gave a strong clue to the propulsion system. A smoke ring strongly suggested a type of gas turbine propulsion system much like I was discovering in my research.

In fact, I felt the major reason someone was trying to get Heflin's photographs from him was because of this clue. Although Mrs. Wells did not reply to that, I pointed out the resemblance to Adamski's Scout Ship photographs to those of Heflin and George Stock's UFO photographs taken in

New Jersey in 1952. But, still no reply from Mrs. Wells.

I tried again by mentioning a UFO photograph taken on November 23, 1951 by Guy B. Marquand, Jr., near Riverside, California, and that it resembled the other UFO photographs. At that statement, Mrs. Wells spoke up and advised me to drop my UFO research efforts altogether as I would get nothing for my efforts than more trouble. She would not elaborate further and completely ignored my comments about the mysterious voice incident. So we said goodbye and I left.

It should be noted that Guy B. Marquand, Jr., stated publicly that the UFO photograph credited to him was a hoax which he perpetrated. He said it was because he was a young man having some fun. He would, however, not say how he pulled the hoax. My strong belief is that he had been told to discredit his own photograph and was merely following orders.

When I returned from my California visit, I moved back to Washington state. I had decided to abandon my interest in UFOs which had nearly become an obsession. I got rid of nearly all the UFO materials and books. I kept only the ones of greatest importance.

I stopped discussing UFOs and tried not to have anything to do with the subject. For awhile, it seemed to be working. Although I still received mail about UFOs, I resolved not to answer any of it. I really thought my days of UFOs and research were over and I was happy about it. Little did I know that this was just the lull before the storm.

One night, I saw a television program narrated by Walter Cronkite about UFOs. It was a tongue-in-cheek presentation and he minimized the Adamski photographs and incident. I became upset, knowing this was a deliberate attempt to portray the UFO information as false. I wrote to Mrs. Wells, told her about the television program, and sent information about ancient symbols, including the Caduceus of Mercury.

Two weeks passed with no word from her. Then at 8:00 p.m. one night, someone knocked at the door. A tall, bearded, black man stood there, neatly dressed in a business suit, wearing a top coat. He said he was selling maga-

34

zines. Although I was not in the magazine-buying mood, I was intrigued when I saw the magazines in his hand, very old *Liberty* magazines which were no longer in print as well as other early 1940s magazines which I hadn't seen since I was a boy. However, they *appeared* new. As we were talking, my family came home. I knew I couldn't talk about UFOs with them there, so I suggested that we talk later. He nodded and left.

I felt terrible about it but if I had started talking about UFOs, I would have been in serious trouble with my family.

About a week later, I received a card in the mail with no name on it, just a Seattle postmark and a drawing of the Adamski UFO Scout Ship on the card with the following quotation, "Be not forgetful to entertain strangers for there by some have entertained angels unawares." (Hebrews 13, 1-2). I felt crushed and deeply confused.

I knew it was game time again!

CHAPTER SEVEN: Government Credibility

Much has been said about George Adamski being a con-man. It should be obvious to those open-minded individuals that evidence points to the fact that Adamski was truthfully involved in the UFO mystery. It seems that any researcher who tries to shed light on these truths runs into trouble in doing so.

One outstanding flaw, the fact that the American public takes things at face value and makes hasty judgments, has been used by the government and others to cause confusion and chaos in the minds of that very American public. It is necessary for the public, news media, and Congress to take those steps required to remedy the situation.

Patience, persistence, and perseverance must be the way to find the truth.

As one who has had some UFO encounters, it appears that many UFO researchers ask the wrong people the right questions and the right people the wrong questions. I recommend that inexperienced UFO researchers ask more questions of the persons who have had UFO encounters instead of so-called UFO experts who have gained their knowledge from unreliable and secondhand sources.

The only true UFO experts are the intelligent beings who command the mysterious machine themselves. Others can only be regarded as serious pioneer researchers or students of the subject. It is important to know that from the late 1940s on it has been the ordinary lay person who drew public attention to the UFO phenomenon. Currently, it is still the lay person who furnishes many of the facts and information in the field of UFOs. The public needs scientists and the scientists need the public. However, many times the lay person is the better source of information.

The American taxpayer would do well to ask himself why the USAF chose an astronomer, J. Allen Hynek, to act as public spokesman about UFOs and Swamp Gas. The Air Force, a government agency, deals with air and space craft of all kinds. We, as American taxpayers, should demand that we be told the truth about UFOs. Does it not seem odd that the Air Force, having so many professionals in the business of identifying air and spacecraft in the process of protecting our country, should suddenly call in astronomers, such as Hynek and Menzel, to identify the mysterious Flying Saucers? Why didn't the Air Force produce some of their professional pilots who were ordered to pursue UFOs and let these experts act as spokesmen? Could it be that the government wanted to perpetuate the idea that UFOs were indeed extraterrestrial? If this is the case, it would explain why Hynek and Menzel were chosen to be the UFO public relations people.

Why did not the federal government bring Adamski to trial for mail fraud if he indeed was lying about his UFO encounters and selling fake photographs and other false UFO information through the U.S. mails? Why was not the persons who supposedly faked the Straith letter on official State Department stationery and mailed it to Adamski through the U.S. mail in order to discredit him in some way, not taken to federal court for this obvious violation of federal law? It is clear that the U.S. government didn't want Adamski's case brought to court on any account. Why? If we had open Congressional hearings on UFOs, Congress could prove to the public one way or the other if it is true that Adamski traveled with a passport bearing special privileges and if so, why?

Many persons, professional and laymen alike, rely on hypnotism and lie detectors as reliable tools in researching UFO encounter stories. If these are reliable tools, why did none of Adamski's critics suggest that he submit to one of these tests? Adamski has fantastically clear UFO photographs and a good deal of technical information and nobody thought to check out his reliability in this way. Why?

Everyone is entitled to their own opinion about UFOs but if Congress would hold open hearings on UFOs, the public would be able to discover the truth about the rumor of an engineer who is employed in the publishing field while also being employed by the federal government as a full-time

"UFO basher."

Where does anyone get the right to practice so much character assassination under the guise of legitimate UFO research?

Could it be that the government itself, in some cases, spent taxpayers' money to manufacture its own UFO stories of encounters with little men to help confuse the UFO mystery even further for the public?

The UFO phenomena is one of the biggest mysteries in the history of mankind. Because UFOs are mysterious, researchers are bound to make mistakes when dealing with the unknown. But mistakes often times lead one to the right answers.

The information within this book is VALUABLE. In time, someone, somewhere will use the information herein and will succeed in constructing an aircraft capable of duplicating many of the operational feats credited to UFOs or flying saucers.

Given time the foregoing will become fact in spite of anything anyone tries to do to prevent it.

The Adamski UFO case appears to have been part of a public educational program in an effort to inform the world public that mankind as EARTH people know it, is not alone.

The Adamski type UFO scout ship is a model -T, simple enough for us to duplicate and use as a tool for learning etc., as it was meant for us to do. Adamski did his part and I have tried to do mine.

Government bureaucrats along with civilian fast buck artists have had it their way for too long in keeping the public confused and in the dark about UFOs in general and the Adamski UFO case in particular. It's time the public put a stop to it. This book will help bring that about.

CHAPTER EIGHT: UFOs and Ma Bell

In order to protect my UFO research work, I have explored several possibilities. I made inquiries about applying for patents on my propulsion system. This led to legal, financial and time problems. Also, I was researching the craft's propulsion design (Adamski's UFO Scout Ship) which someone had already perfected. There is evidence in ancient Indian Sanskrit that intelligent beings had designed and manufactured a similar propulsion system. Patent lawyers told me there may be more than one patent involved and it could be stymied because of the laws of national security. In other words, if my patent were accepted, the government could deem it detrimental to national security and hold it for an indefinite time. They could also order me to keep my findings from the public under penalty of fine, imprisonment or both plus losing any inventor's rights I may have. So, struggling along, I have taken what steps I could to protect my work.

However, since I needed a working model for public demonstration and didn't have the money to build it, I had to find a way to interest the public. Since the public has been conditioned to view any circular aircraft design with three globular units on the bottom side as a fake, the situation seemed hopeless.

There was, indeed, someone unknown interested in my work. In a letter from a friend in Oregon, October 31, 1966, he mentioned a phone conversation between the two of us a few nights earlier. "Bill, I hate to alarm you, and I think you know me well enough to know that I'm not joking. Right after you hung up, the phone didn't go dead. I heard a man's voice say, 'Ward, did you?' Then the line went blank. There wasn't any dial tone until after I hung up." I didn't know what to think.

Over a long period of time, I had problems with telephones.

I wanted to raise money to produce the small working model so I contacted Ray Palmer again with the idea of publishing a book about my UFO experiences as he had shown interest in it. We talked frequently on the phone and this led to an article in the February 1969 issue of *Flying Saucers* magazine.

Although my phone bills were high, they were paid so I didn't understand the phone company's nervousness about them.

I also didn't understand why the phone company contacted Ray Palmer and asked him exactly what our discussions were about UFOs. Before Ray told them it was none of their business, he got a name and passed this information on to me. I contacted the phone company and they admitted what had happened but I couldn't learn who ordered this action or why it was ordered.

I do not hold every agency of the United States government at fault for denying the existence of UFOs to the public. But I do hold some of these agencies and their employees responsible for having spent taxpayers money to withhold the truth about the existence of the UFOs from the public. I think these actions over the past forty years in this matter is criminal. I believe the United states Congress should obey the will of the people and do something about it to the satisfaction of the American public.

The late Dr. J.E. McDonald, a University of Arizona professor, attacked the United States government policy on UFOs as "badly mishandled." I agree with him. At his lecture in Seattle in 1968, he stated that the Adamski UFO case was not worth discussing. I strongly disagreed with him on that, however. I did agree with him again when he made the following statement, "Current scientific attitudes toward the UFO problem must be radically alerted."

I felt I had to bring attention to this coverup in some way so I wrote to the National Enquirer and expressed my concern about the news media and government coverup of the UFO phenomenon. I felt it was a place to start since the National Enquirer was a national tabloid read widely. I received the following reply on December 14, 1968.

NATIONAL

ENQUIRER

210 SYLVAN AVENUE
ENGLEWOOD CLIFFS, N. J. 07632
569-5600

EXECUTIVE OFFICES

December 14, 1968

Mr. Bill Clendenon
P. O. Box 49
Port Angeles, Washington 98362

Dear Mr. Clendenon:

Thank you for your kind comments on the UFO
articles published in The ENQUIRER.

I agree that much of the news media prefers to scoff
at this problem -- but for what reasons, I can't fathom.
Through the countless articles we have published on this
fascinating subject, we have tried to bring pressure upon
the government to reveal the truth to the public. We feel
that the public is fed up with the ridiculous answers provided by the government to explain UFO sightings. We still
hope we can someday succeed in forcing the government to be
open and honest on this problem.

I have turned your letter over to the UFO expert on our
staff and if he has any questions or feels your particular
case is worth pursuing, he will contact you.

Rest assured we will continue publishing articles on this
topic. I hope you find them of interest. Please let us know
if you do. We're always happy to hear from our readers -- for
you're the people we're trying to please.

Yours truly,

NAT CHRZAN
Editor

NC:hg

41

CHAPTER NINE: Friction

In 1968, I began receiving mail from Charlotte Blobe of Clintonville, Wisconsin, and Thomas Heiman of Alexandria, Virginia. Later, I received mail from these two people from the same address in Valley Center, California, in 1974. I was still receiving mail from Alice K. Wells. Blobe and Heiman implied that I should deal with them in regard to my research program of George Adamski and UFOs.

Meanwhile, I learned that there were UFO films taken by Adamski and Mrs. Rodeffer before Adamski's death. Since I had not seen any of these, I was trying to obtain copies for photo research and comparison with other UFO photos. In 1966, I wrote to Mrs. Rodeffer but received no answer. I also wrote to Blobe but received no answer.

At this time, I was moving back to the eastern part of the United States. I decided to stop in Vista, California, to visit Adamski's home to talk with Alice Wells one more time. I noticed things in Adamski's home were markedly different than the last time I had visited. All the UFO photos, models, and paintings were gone. When I asked Mrs. Wells where they were, she replied that she had fallen asleep with a cigarette, the house had caught fire, and all these things were destroyed. As I looked around me, I had to wonder how all of the UFO items were destroyed without damage to the house and Mrs. Wells with it. I kept my silence.

I asked Mrs. Wells if she could let me use some of the UFO photos taken in Silver Springs by Adamski and Mrs. Rodeffer. She quickly referred me to Mrs. Rodeffer or Mrs. Blobe. This confused me for previously Mrs. Wells had indicated that there was a great deal of friction between her and Mrs. Blobe, but now she was telling me to go and see Mrs. Blobe. I mentioned to Mrs. Wells that I had a letter from Mrs. Blobe inviting me to stop in if I came to California.

Although I knew Mrs. Wells to be an intelligent woman, she could don a cloak of naivete which left one confused and uncertain of her meaning. I had seen this before and it would happen in the future, too. (Looking back, I believe that Adamski and Mrs. Wells were being advised by persons in authority.)

Saying goodbye to Mrs. Wells, I headed for Valley Center to visit Mrs. Blobe. When I arrived at her house, there were many cars parked outside. I drove in the driveway and noticed a man and woman watching my every move. I sat in my truck for a few moments, trying to decide just what to do. As I approached the house, the woman said something to the man and left. I stopped where I was and let the man approach me. He introduced himself as Mr. Heiman. I explained to him who I was and said I had a letter from Mrs. Blobe inviting me to stop in when I visited California.

He curtly told me that Mrs. Blobe was gone for several days. It was plain that he would not give me any information about Mrs. Blobe, UFO photographs, or anything else. He did not invite me into the house. I had a strong hunch that he and Mrs. Blobe knew who I was when I drove in the driveway and I am convinced they knew I had visited Mrs. Wells first.

This, of course, made me wonder about the public display of friction between Mrs. Wells and Mrs. Blobe. Was it all an act? And, if so, for what reason? Also, why would Mrs. Blobe invite me to her house and then pretend to be gone when I got there?

CHAPTER TEN: Mysterious Events

After I returned to the eastern United States, I had more UFO night sightings, but only one daytime sighting with witnesses. Most of these nighttime sightings were colored balls of light. The nighttime sightings were in Tennessee and the colored balls of light were orange, white, and the familiar greenish color others have seen. All of these UFOs flew on a horizontal course with no sound. However, in the case of one of the green ones, I was in for quite a show. It was 8:00 p.m. with lots of stars visible as I was sitting in the backyard looking upward. Soon, I heard the sound of propeller driven aircraft approaching on a westward course. There were five military planes in a somewhat loose V-formation at an altitude where their navigation lights were very visible. Suddenly, the ball of green light approached the plane formation on a head-on horizontal collision course from the west. It looked like an aerial bowling game for the UFO headed straight into the five plane group. Immediately, the planes took evasive action by scattering out in all directions to avoid the UFO. I could hear the noise pitch of the plane engines change due to their evasive maneuvers. It appeared that the UFOs action was deliberate as there was time for it to change course. It also appeared to me that the UFO was trying to draw attention to itself.

By now I had learned my lesson about reporting UFO sightings to the authorities and made no effort to do so.

The daytime sighting also occurred in Tennessee one afternoon when I was leaving a shopping center. While putting groceries in my car, I heard people yelling and pointing skyward near the railroad tracks that ran east and west parallel to the shopping center. The UFO was an orange ball of light and flew horizontally at a high rate of speed. There was no sound.

Over the years, I received mail or items from unknown sources or which originated in ways I could not verify. For instance, I received a copy of Lenord G. Cramps UFO book, *Space, Gravity, and the Flying Saucer,* in such a

Letter from Unknown Source

Dear Mr. Clendenon:

The description of the motive device for your vehicle is most interesting.

The device which you are trying to build must be circular because it must contain a large rotating part as its primary motive principle.

A strong electrical or magnetic effect must be evidenced.

These requirements might best be met by the following arrangement of parts.

Arrange powerful magnets, norths facing out, along the rim of wheel. Rotate the wheel to the right. A force may be generated opposite that of gravity.

The vector equation for this phenomenon is ideally

$$F = v_r \times H$$

in which F is the force directed away from the gravitational force, v_r is the rim bvelocity, and H is the magnetic field strength in gauss.

Working out the equations for the vector multiplication, I found that the quantity mass-squared is "left over". I take this to mean that the device should be less efficient as the square of the mass increases.

Using 10 1000-gauss magnets and a rim velocity of 100 cm/sec. with a mass of 1000 grams, the device should generate an upward force of 1 dyne. This is very small, but can be increased by increasing the power of the magnets, and the rim velocity.:

$$F = v \cdot H/ m^2$$

Good luck. By the way, this device may generate radiation of a rather characteristic nature, so that you may have visitors

Envelope Postmarked Midwestern College Town

manner. I received unsigned notes or letters of a technical nature. One time, Alice Wells called me long distance and asked me to talk to a gentleman whom she did not identify. The man asked me a number of technical questions about my propulsion research efforts. These things spurred me on in my propulsion research.

Alice K. Wells kept in contact with me over the years but her reactions and comments were contradictory and left me confused. However, I kept experimenting in every way I could to find a way to produce a working model of the propulsion system.

Civilian UFO groups were not much help because of their conservative attitude toward George Adamski so I

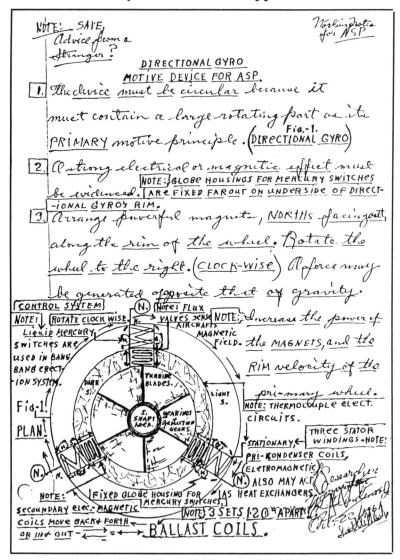

struggled on alone as best as I could.

Wendelle C. Stevens, Lt. Col., U.S.A.F. (Ret.), well-known in the UFO research field, published an article in 1975 about bell-shaped UFOs in *Official UFO* magazine. In this article, Wendelle included information about my sighting of an Adamski-type UFO Scout Ship which is bell-shaped. Through the years, I was to have contact with this man several times. I also was in touch with an editor of this magazine, Dr. Dennis William Hauck, a mathematician and author, also well-known in the UFO field of research.

In 1977, I contacted Phillip Bergman, Senior Producer for ABC's Saturday Evening News with Ted Koppel in New York, regarding a story on UFOs. One of the pieces of correspondence which came from him is as follows:

ABC News   7 West 66th Street   New York, New York 10023   Telephone 212 LT1-7777   **- EXTENSION - 5115.**

Dear Mr. Clendenon:

Thank you for your story suggestion. It has been assigned
to a Producer, who is looking into it. Your suggestion
could very well be the basis for a Closeup spot on the
Saturday Evening News in the near future.

Thanks again and if you have any other suggestions, please
send them to us.

Sincerely,

Philip Bergman
Senior Producer
Saturday Evening News
with Ted Koppel

Mr. W. D. Clendenon
P.O. Box 1961
Fort Walton Beach, FL 32548

April 11, 1977

SECRETARY - GLORIA CURRY,

At the same time I was in touch with Dr. Dennis Hauck as he had indicated his interest in doing an article about my UFO research in *Official UFO* magazine. This magazine seemed to present a serious attitude toward all sides of the UFO mystery so I was anxious to work with Dr. Hauck.

I put Dr. Hauck in touch with ABC News and he contacted CBS News in Washington, D.C. I understood that the final result was to be a television program which would include the George Adamski case. Therefore, I suggested to Dr. Hauck that he contact Alice K. Wells. He invited her to participate, and everything seemed to be working well.

However, a series of events took place in a short period of time which appeared too coincidental to be a coincidence.

One, Dr. Hauck notified me that all the plans for the publications or television programs having anything to do with Adamski had been canceled.

Two, Dr. Hauck was fired as editor of *Official UFO magazine.*

Three, for some reason the remaining staff of *Official UFO magazine* began to discredit their own publication by publishing silly photographs and comments.

Four, *Official UFO magazine* ceased publication.

Five, I received a phone call from a person in Washington, D.C., who said he was employed by CBS. He said there would be no television show about George Adamski at this time although it could be considered in the future.

All of the actions seemed to be more than coincidences. Although I have tried, I have been unsuccessful in contacting Dr. Hauck again.

The cancellations of the television programs and publications Dr. Hauck and I had been working on made me feel helpless. I had put in years of effort and did not want it to be in vain.

I decided to go to Washington, D.C. and see if I could accomplish something in person. I drove straight through to Washington and arrived the next day after experiencing trouble with my car. Having spent most of my funds for repairs, I felt that my trip would be shortened considerably so I set out to do the most I could with the little time I had. I went to a telephone located in a park across the street from the White House. President Carter had said he had seen UFOs and if elected President, he would help the American people get the truth about the UFOs existence.

UFO cover-up

Whatever happened to President Carter's electoral promise that he would shed all the light that needed shedding on the subject of UFOs?

It was an important promise, if only because there are many people who believe in flying saucers, though I hasten to point out I don't.

But Jimmy Carter says he does. And he said that if he was elected, the American people would finally be told the real story on UFOs.

Jimmy claims he saw his UFO when he was in Leary, a small town in Georgia. The President described the thing he saw as the size of the full moon. He also said that it changed color.

According to his press secretary Jody Powell, the President was sure it was a UFO. Said Powell:

"I remember Jimmy saying he did in fact see a strange light or object at night which did not appear to be a star or planet or anything he could explain."

It's been six months now and we're still waiting for the President's explanations.

UFO believers were heartened to learn that their new President, like them, believed we are not alone in the universe. And yet after Carter's revelation that he too had seen a UFO, there's been nothing further from the man.

Official voices, in fact, have said that Carter never did see a UFO; it was a star or — why not? — swamp gas.

So what's the story, Mr President? Have you too joined the great conspiracy of silence that prevails within Government on the subject of UFOs?

I think it's time for some answers. And there are millions of voters who think likewise.

Turn to pages 4 and 5 for "UFO Flie."

An excerpt of Mike Dorland's column which appeared in 1977.

49

Of course, I was not allowed to talk to President Carter. Instead, I was shunted to one of his aides, an impolite man who would not give me his name. He told me, ''If you want something that flies, get a bird.'' He also told me to go back where I came from before I got into trouble. So much for presidential promises on UFOs.

I returned to Tennessee in September, 1977 and was contacted by Mike Dorland, the writer and reporter for the National Examiner from Montreal, Canada. He was publishing a series of heated articles about President Carter's failure to force the truth about the UFOs existence out in the open. We agreed to work together on another series of articles about UFOs in general and George Adamski's case in particular. Mike said the editor had given him the go-ahead sign and everything was ready. I sent my material to Mike and received this reply:

E *NATIONAL*
EXAMINER

1440 ST. CATHERINE W., SUITE 625, MONTREAL, QUEBEC, CANADA, H3G 1S2. Phone (514) 866-7744

September 20, 1977.

Mr W.D. Clendenon
PO Box 63
Morrison, Tenn.
37357.

Dear Bill,

   A mountain of material just landed on my desk, so this is to acknowledge that I received it all right.

   I haven't yet been able to do more than glance at it, but it appears most interesting.

   Who is Alice K. Wells and who are the Brothers? You did not mention them earlier. Where are they located and what do they do?

   I trust your health is bearing up and that, if you had your operation, it came off well.

   I'm working on my editor to try and run your story this week. As I said earlier, I'll send you all copies as soon as they're printed, as well as one to Sen. Howard Baker.

   Please keep in touch.

                              Sincerely,

                              Michael Dorland
                              Michael Dorland

The Brothers? Alice K. Wells? What could I tell him about her? The reason I mentioned the Brothers to him was because of the letters I had received from Alice K. Wells which are as follows:

February 19,1977

Dear Bill Clendenon.

  Thank you for all you are doing for the Brother's work.
I feel that George Adamski and the others are very aware of what you
are doing, and are guiding you in what to say.

  The material that you sent is interesting. Fred Steckling x
will be up tomorrow and I will give it to him. He is a very busy
man but will send you material that you can use, as soon as possible.,

  Please accept the enclosed check - a small token of appre-
ciation for all that you are doing.

Most sincerely

Alice K. Wells -

September 1,1977

Dear Bill Clendenon.

  The photostats received. Two of our represenative
were at the Mexico Fiesco-- It was just that, poorly
arranged, manage and the same old Garbage peddled
by the prophets of GLOOM AND FALSE REPORTS.

I would suggest that for your own health and well
being- that you forget about the whole UFO saga, as
presented to the people.

The Brothers kmow what these negative mongers are
doing and they also know that their claims have no
substance and in time they will deteriate with the
Junck they are peddling.

People who are looking for Truth and a Better Way of
Life, are not interested in their prattle.

So, do not consern yourself, look on the bright,
constructive, lasting side of that which is ETERNAL.

If you will do this -- you will be much happier and
those about you will detect the change and enjoy
your company.

I trust that you will see the wisdom of this advice
that comes from the Brothers.

                    Most Sincerely

                    Alice K. Wells --
                    Alice K. Wells

Again, coincidence reared its head on the happenings concerning my research!

One, Mike Dorland was not able to publish the Adamski UFO articles and he never gave me an explanation.

Two, Mike Dorland was fired from the National Examiner.

Three, I have never been able to contact Mike Dorland and I have made plenty of attempts to do so.

What can I attribute all these coincidences to? Are they coincidences? I don't think so.

NOTE:
If one wants the general public to be informed on a subject quickly, the Supermarket tabloids are not to be ignored. Once in awhile they do get off sex and print information that can be regarded as a valuable public service. With all the money being made on so many far-out UFO stories today, I, for one, cannot understand why the tabloids, including the National Enquirer and especially the National Examiner have almost come to a complete halt when it comes to publishing UFO stories of any kind. Could it be that somewhere along the way over the last 40 + years their stories about UFOs, the Hollow Earth, and yes, even the UFO case of George Adamski was beginning to draw too much public attention where it wasn't wanted? In their case has UFO come to mean Uncle Frowns On it?

DEFENCE RESEARCH BOARD

DEPARTMENT OF NATIONAL DEFENCE
CANADA

Ottawa 4, Ontario,
March 23, 1966.

Mr. William D. Clendenon, Jr.,
P. O. Box 926,
Portland, Oregon, U.S.A.

Dear Sir:

Your letter of March 11 has been passed to the
Defence Research Board for reply.

We appreciate receiving your views on the subject
of UFOs but feel you should be informed that Canada is not
carrying out investigations in this field.

About 10 years ago, a federal government committee
was established to investigate sightings of UFOs reported in
Canada. Virtually all sightings were unquestionably related
to weather balloons, aircraft, falling stars or other
identifiable phenomena. This committee has been inactive for
many years.

Yours truly,

(C.A. Pope)
Information Officer.

DSM—11-61 (96 8795)

CHAPTER ELEVEN: International Research

There has been much controversy about *Pioneers of Space,* a work of fiction, one of the first books George Adamski wrote. This book, published in 1949, was copyrighted under the name of *Professor* George Adamski several years before he gained fame as a UFO contactee. Adamski was not a professor in any sense of the word, but he told me that *Professor* was his nickname and he used it as a pen name. The term *professor* was later used against him. The use of pen names is common among fiction writers and is not illegal. In all fairness, though, I agree with other UFO researchers that the contents of *Pioneers of Space* is very close to the contents of *Inside the Space Ships* and other UFO works Adamski was to publish at a later date. But it has also been pointed out by others that the possibility is there that the contactors read *Pioneers of Space* and decided that Adamski be recruited to play in a fantastic UFO cover story, with or without his full knowledge of the facts. This hypothesis could have some interesting possibilities should one take the time to sift through the clues.

In Adamski's *Flying Saucers Have Landed* (1953) there is a photograph opposite page 113 (plate #8) which describes writing from another planet. The strange writing when deciphered is thought to be a brief technical account of how the Adamski-type UFO Scout Ship is propelled. Adamski said the photo-plate was given to him by a UFO Scout Ship crew in 1952.

It is interesting to note that Professor Marcel Homet, an archaeologist known worldwide, claims to have discovered ancient Indian petroglyphs that appear almost identical to the mysterious writings on photo-plate #8.

In 1974, I received the following information from persons said to be well-acquainted with George Adamski. In 1963, internationally known Professor Marcel Homet published his research findings about ancient civilizations in

Peru and Brazil in his book, *Sons of the Sun*. He claimed to have discovered hundreds of remarkable petroglyphs which he believed to have been made approximately 20,000 years ago by a race of people who were in contact with extraterrestrials. The Professor was acquainted with Adamski and was Said to have made the following statement in Adamski's behalf, "I have never found Mr. Adamski to be untruthful."

When the Homet research information is compared with the research of Desmond Leslie and James Churchward into ancient Sanskrit records of India, the possibility of Adamski being part of a fantastic UFO cover story gains more strength. However, this demands more research. There is plenty of work for those who are serious about researching UFO background information.

The research work of the late Dr. Ivan Sanderson was similar to the foregoing, for as science editor for *Argosy*, he wrote an article in November 1969 about an ancient Columbian artifact which resembled modern jet aircraft. Dr. Sanderson and I were collaborating on an article for *Argosy* magazine on ancient Indian power packs for aerial propulsion which included the liquid metal mercury. But Mr. Sanderson died before this article was a reality.

Dr. Khalil Messiha, Egyptologist and prominent archaeologist, discovered an ancient artifact stored in the Cairo Museum on July 14, 1969 bearing strong resemblance to a modern aircraft or model glider. Dr. Messiha and Dr. Sanderson pointed out that ancients of Egypt and India traded goods. Therefore, it was possible that they also exchanged information in the science of aeronautics.

The excellent research efforts of British UFO researcher, Leonard G. Cramp, an aeronautical engineer, on the UFO photographs and technical information furnished to him by George Adamski had been very helpful in my own UFO research program. Cramp's comparison by orthographic projection of the Adamski and English Coniston UFO photographs are excellent examples of the proper scientific analysis that can and should be carried out on information that George Adamski furnished to the public. The dimensions for the Adamski UFO Scout Ship as described by Mr. Cramp's research appear to be nearly correct from my own close observation of the Adamski-type UFO Scout Ship. It may also be true that the three-ball undercarriage of the Adamski UFO acts as buoyancy chambers in case of an emergency water landing. Mr. Cramp's two excellent reference books on UFOs, *Space, Gravity, and The Flying Saucer* and *Piece for a Jigsaw* are very valuable tools for the serious UFO researcher. According to my own observations of the UFO aircraft as Mr. Cramp's drawings describe on page 176 of *Space, Gravity, and The Flying Saucer,* I will have to disagree that any part of the power plant extends below the base of the outer flange of the UFO Scout Ship at any time for the following reasons:

1) the outer flange, as I see it, must be stationary to act as a support for the whole ship while it rests on any surface.

2) the power plant has to be clear in order to revolve or run, thus the outer flange, while acting as a support overall, also allows running clearance for the turbine of the ship while at rest on any surface.

Mr. Cramp deserves a great deal of credit and more recognition for the research he had done in the case of George Adamski.

On February 26, 1965, Madeline Rodeffer and George Adamski took a color movie film of a UFO at Silver Springs, Maryland. However, the film had a more negative than positive effect. There is some evidence that the film was sabotaged in some manner during processing. This is the UFO film which Alice K. Wells kept me from viewing in 1975. In spite of her efforts, I was finally able to view the movie and obtain still copies from it, some of which were provided by Mrs. Rodeffer again. It occurred to me that the particular mechanical movement of the three ball undercarriage of the UFO was technically incorrect.

Therefore, the possibility exists that the mechanical movement of the three ball undercarriage of the UFO is deliberately incorrect to throw off proper technical analysis of the UFO in the film and thus further discredit George Adamski.

But, if this was done, was it done with or without his knowledge? If it was done with his knowledge, it is my belief that he was following orders from persons unknown.

Regarding the movement of the three ball undercarriage of the ship, I believe the public is expected to view this up and down mechanical movement of the three ball undercarriage on the UFO as an old-fashioned centrifugal governor movement. I believe the UFO film was sabotaged. I believe that the unknown persons were fearful that the Adamski UFO cover story was in danger of being exposed. This exposure would show the true origins of the UFO Scout Ship shown in the 1952 photographs.

When researching something with the magnitude of the UFO phenomenon, all possibilities must be considered.

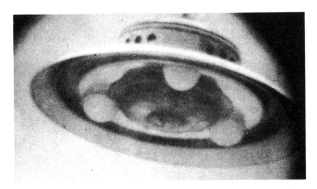

The above famous photo was allegedly taken by George Adamski in the early '50s. Among other things Adamski claims to have gone to the moon and to have traveled from Kansas City to Davenport, Iowa in this ship. It has been noted that Adamski's saucer looks strikingly like a chicken brooder shown in the lower drawing. It was also assumed that Adamski had photographed bottle cooler lids that had exactly the same design as Adamski's saucers. However, it was discovered that the bottle cooler's designer had designed the bottle cooler lid in 1959 six years after Adamski published his photographs. In fact, it turned out the designer was a follower of the UFO mystery and his design was a tribute to George Adamski! It's also worth noting that Adamski claimed that the hemispherical landing pods located on the underside of the craft were extendable landing gear. Assuming that the photograph is genuine, those pods would more likely be guidance and field stabilization components (vortex pods). Photographs courtesy of Openhead Press, London, UK.

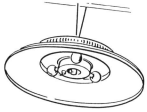

BOTTLE COOLER LID   CHICKEN BROODER

UFO RESEARCH

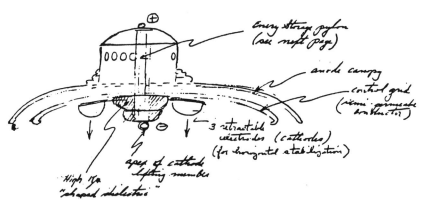

NOTE: Page from notebook of T.T. Brown, etc.

T. Townsend Brown

In 1957, T.T. Brown and Associates met with George Adamski. The purpose was to **LEARN WHAT THEY COULD ABOUT THE DE-SIGN** of the UFO or Venusian Scout Ship, photographed by Adamski in 1952. In reference to the research, refer to pages 206, 207 and 208.

THOMAS TOWNSEND BROWN, Physicist, Inventor
1905-1985

- 1956 - T. T. Brown formed UFO Study Group, NICAP, in Washington, D.C.
- 1957-1960 - T.T. Brown hired as a consultant for Whitehall-Rand Project under Auspices of Bahnson Labs (Agnew Bahnson) to do antigravity R. & D.
- 1958 - T.T. Brown formed Rand International, Ltd.

58

CHAPTER TWELVE: The Important Five

In the past 38 years, five Americans have photographed UFOs, producing photographs which have the power to unlock the mystery of the UFO phenomenon if analyzed properly. Those men are:

1. *Guy B. Marquand, Jr.* photographed a UFO in flight near Riverside, California, on November 23, 1951. Although it caused a nationwide sensation at the time, he now says his UFO photograph was a hoax but declines to say how he made the photograph. This leads me to ask: Was Marquand persuaded by persons unknown to discredit his own work? If this is the case, who and why?

2. *George Stock,* Passaic, New Jersey, made a series of UFO photographs on July 28, 1952. The United States Air Force analyzed newspaper reprints of the photographs and not the originals. The UFO was in sight for seven minutes at 10:15 a.m. but no other reports of the sighting existed. The USAF stated that by comparing the sizes of the objects in the foreground, they determined the UFO was the size of a ''lady's sun hat,'' thereby discrediting it.

I would like to know why the USAF did not contact Mr. Stock to enlist his cooperation in the proper analysis of the originals. The photographs were taken in broad daylight, at close range, with a common camera. My own experience with reporting UFO sightings has taught me the futility of this. It is not surprising that these events occurred.

3. *George Adamski* photographed a UFO which he termed a Venusian Scout Ship at 9:10 a.m. on December 13, 1952 at Palomar Gardens, California. At a press conference in Detroit, Michigan, on March 25, 1966, the news media asked Dr. J. Allen Hynek, USAF, special investigator on UFOs, what the UFOs were. He replied, ''swamp gas.'' Hynek was also asked about the UFO in the photograph taken by Adamski in 1952. At that time, he replied that it was a ''chicken feeder.'' In an April 20, 1966 letter to Senator Wayne Morse of Oregon, the USAF stated that this

photograph was analyzed by the Air Force and it was determined that it was a "tobacco humidor top and a baby bottle nipple attached to the top and three ping pong balls on the bottom."

Why did the United States government officials come to such different conclusions on the same Adamski UFO photograph?

Why has this been overlooked by the media and civilian UFO research organizations for so long?

With over 40 years research on UFOs, 37 years of that researching the Adamski UFO Scout Ship photograph and as a former United States Navy Aircraft Identification Instructor, I can state, as an eye witness, that the object in the Adamski UFO Scout Ship photograph is an intelligently controlled aircraft, a flying machine of unknown origin.

4. *Rex Heflin,* a Los Angeles County Highway Engineer photographed a UFO near Santa Ana, California, August 3, 1965. After USAF analysis, it was determined that the saucer-shaped object was 36 inches in diameter as opposed to Heflin's claim that it was 30 feet. I agree with the Air Force but only in that the so-called ground disturbance was in actuality dried grass, not a ground surface disturbance caused by the UFO.

Heflin's remarkable photographs are praised by many UFO researchers, as I think they should be. However, many UFO researchers also agree that there is a strong resemblance of the Heflin UFO to the side view illustration by Glenn Passmore as described in diagram #9 of Adamski's book, *Inside the Space Ships,* published in 1955, ten years before Heflin's photographs were taken.

The Passmore illustration was called a Saturnian Scout and said to be a little larger than Adamski's Venusian Scout. Mr. Passmore made the illustration of this UFO from the information given to him by George Adamski who stated that he had been aboard this type of aircraft. There is, of course, the famous smoke ring photograph, one of the four photographs Heflin took. This strongly indicated that this UFO was propelled by a type of air breathing gas turbine propulsion system. Mr. Heflin, an engineer, not only correctly estimated the size of his UFO but his photographs caught the UFO hovering and tilting downward in a wobbly gyroscopic-like motion. When analyzing Heflin's photographs, note the similar shading effect on his UFO when compared to Adamski's and Marquand's. All of the clues point to truth. There is also the reality that a man in a United States Air Force uniform persuaded Heflin to surrender the Polaroid prints. That person knew the value of the prints if they were analyzed properly.

5. *Kurt Kreitz,* Lancaster, Pennsylvania, took two photographs of a UFO in February, 1967. In one of these, the three-ball undercarriage of the ship was there to see. In his photographs, the high power electrical lines and steel towers are visible to help in analysis. The letters at the end of this chapter point out the official "run-around" I was given when I tried to obtain more information about the Kreitz's photographs for my own research. In fact, someone approached me and wanted to have the copies I have of these photographs. The USAF would not comment on them.

It is important to note the dates between Adamski's 1952 UFO Scout photo and those photographs taken by Kreitz in 1967, fifteen years to be exact.

In all my research, I found little criticism of Kreitz's photographs from any source. What stands out is the lack of public acknowledgement by UFO researchers to the strong resemblance of the three-ball undercarriage of the object in both Adamski's 1952 UFO photo and the UFO photo taken by Kreitz in 1967.

There is a strong resemblance to the objects in all five men's photographs. Why hasn't the United States government noticed that? Or have they??

Has the American or even world public been kept in the dark? Why are we not given the truth even when our tax dollars are the revenue which promotes these coverups and confusion. When will the American public stand up and say, "Enough!"?

The following is an assortment of letters which illustrate what I am talking about, the run-around, the evasiveness, the total lack of interest in many cases! You be the judge!

Congress of the United States

Office of the Minority Leader

House of Representatives

Washington, D.C.

March 31, 1966

Mr. William D. Clendenon, Jr.
P.O. Box 926
Portland, Oregon., 97207

Dear Mr. Clendenon:

Thank you for your recent communication endorsing my proposal that
Congress investigate the rash of reported sightings of unidentified
flying objects in southern Michigan and in other parts of the
country.

It is proper for the Federal government to look into a matter
which is causing alarm to the people of our nation as these
sightings have, and it is for this reason that I have called for
the investigation.

Such an inquiry is necessary and wholesome because of the incidents
that have occurred and I assure you that I will continue to press
for this investigation. I want you to know that your specific
comments were helpful to me in this respect, and that your remarks
do not go unheeded. I am aware of other reports such as the UFO
Evidence which pose questions that, like the most recent sightings,
cannot be answered by a few "pat" solutions.

Kindest personal regards.

Sincerely,

G K Ford

Gerald R. Ford, M.C.

DEPARTMENT OF THE AIR FORCE

WASHINGTON

APR 21 1966

OFFICE OF THE SECRETARY

Dear Senator Morse:

APR 2 0 1966

We refer to your further inquiry in behalf of Mr.
William D. Clendenon, Jr., relative to his invention and
unidentified flying objects.

The Adamski photograph referred to in Mr. Clendenon's
letter was analyzed by the Air Force. The object depicted in
the photograph was determined to be a tobacco humidor top
with three ping-pong balls attached to the bottom and a baby
bottle nipple attached to the top.

We hope this information will serve your purpose.

Sincerely,

Attachment

Honorable Wayne Morse

United States Senate

FREDERICK H. FAHRINGER, Col., USAF
Congressional Inquiry Division
Office of Legislative Liaison

62

UNIVERSITY OF COLORADO

BOULDER, COLORADO 80302

DEPARTMENT ?XŻZ XŻX ŻXŻXŻXŻXŻXŻXŻ

1005 JILA Bldg.   11 December 1967

Mr. Bill Clendenon
P. O. Box 49
Port Angeles, Washington 98362

Dear Mr. Clendenon:

  Since we are unable to locate any previous correspondence with
you, or the pictures you say you sent to us, I wonder if you would be
able to give us a most specific date.

  If you can give us any concrete evidence, backed by correspondence
or other materials, of your alleged criticism of the Air Force handling
of sightings we would be very glad to have it. Many people make similar
comments, but refuse to communicate further when we ask for good evidence
of such mishandling.

  If we are able to locate your pictures we will be glad to return
them to you as you request.

        Sincerely,

        (Mrs.) Kathryn Shapley
        Secretary to Dr. Condon

GERALD R. FORD
FIFTH DISTRICT, MICHIGAN

MICHIGAN OFFICE:
425 CHERRY STREET SE.
GRAND RAPIDS
ZIP 49502

Congress of the United States
Office of the Minority Leader
House of Representatives
Washington, D.C.   20515

December 12, 1967

Mr. Bill Clendenon
Post Office Box 49
Port Angeles, Washington 98362

Dear Mr. Clendenon:

Thank you for your recent letter with the attached information
concerning the investigation of unidentified flying objects.

As you may be aware, I have repeatedly recommended that thorough
investigative action be taken and the public be informed concerning
this matter. I had called for an investigation of this matter in
the Spring of 1966. Enclosed are some of the news releases which
I issued at that time.

As you also know, the Air Force has authorized a thorough and
complete investigation of UFOs. This investigation is being conducted at this present time at the University of Colorado under
the chairmanship of Dr. Edward U. Condon and we hope it will provide the public with all the facts we need about UFOs. I am looking

63

forward with all other interested citizens to seeing the results of this important investigation. Enclosed are two news articles concerning this current study.

Kindest regards.

Sincerely,

Gerald Ford

Gerald R. Ford, M.C.

GRF:h

Encl.

UNIVERSITY OF COLORADO

BOULDER, COLORADO 80302

DEPARTMENT OF PHYSICS AND ASTROPHYSICS 1005 JILA Bldg.

17 January 1968

Mr. Bill Clendenon
P. O. Box 49
Port Angeles, Washington 98362

Dear Mr. Clendenon:

Returned herewith are the pictures which you sent to this office earlier and which you requested returned on December 7. Sorry it has taken us so long to locate them and get them back to you.

Sincerely,

Kathryn Shapley

(Mrs.) Kathryn Shapley

Returned herewith:
Envelope with 6 photos and
   newspaper clipping
and 11 pages of drawings

OFFICE OF THE SECRETARY

JAN 2 6 1968

Dear Mr. Meeds:

This is in reply to your request for alleged UFO photographs taken by George Adamski, George Stock, and Rex Heflin, and the official Air Force analysis of each.

None of the above-mentioned photographs were ever submitted to the Air Force for analysis; therefore, we are unable to provide the requested copies. However, because of the extensive public interest, informal analysis of each of them has been made from prints that appeared in newspapers and other media.

In the case of Mr. Adamski's photographs, photo analysts at Wright-Patterson AFB, Ohio, have determined that these prints contain stimuli caused by a tobacco humidor and three ping pong balls.

The photographs taken by Mr. Stock were not believed to be authentic. They were judged by the Aerospace Technical Intelligence Center to be taken at 10:15 a.m., and the UFO was apparently visible for seven minutes; yet no one else reported seeing it. In addition, the relative sizes of the items in the foreground indicate the object would be about the size of a lady's sun hat at 30 or 40 feet.

With reference to the photographs of Mr. Heflin, photogrammetry of the print indicated the saucer-shaped UFO in the picture had a maximum diameter of 36 inches, as opposed to Mr. Heflin's claim that it had a diameter of 30 feet. The so-called disturbance in the picture was dried grass in the proximity of concrete protrusions that can be found along that section of the road; it also appears in portions of the photograph other than those contiguous to the saucer-shaped object.

We are returning without comment the picture you sent us since its properties are so poor as to preclude valid analysis. There is no record of a copy ever having been submitted to the Air Force. For your information, we are including an account of it as reported in the November issue of McLean's magazine, which is published in Canada.

Whenever we can be of assistance in any way, you will find us glad to cooperate.

Sincerely,

Attachment

Honorable Lloyd Meeds

House of Representatives

B. M. ETTENSON, Colonel, USAF
Congressional Inquiry Division
Office of Legislative Liaison

65

NO. 2 · A FAWCETT PUBLICATION
75¢

The NEW Report On
FLYING SAUCERS

By The Publishers of *TRUE*

All-New Exclusive Photos & Sightings Direct From U.S. Air Force Project Blue Book Files

DO SAUCERS COME FROM MARS?
By Dr. Charles H. Smiley, Chairman
Dep't of Astronomy, Brown University

Interviews with Maj. Hector Quintanilla,
Project Blue Book Chief;
Long John Nebel

Statements by Wernher von Braun, Marshall Space Flight Center, NASA;
Dr. J. Allen Hynek, Dearborn Observatory, Northwestern University;
John Fuller, author of "Incident at Exeter";
Maj. George W. Ogles, Office of Information, Secretary of Air Force

Cover photograph by
Kurt Kreitz of Lancaster, PA,
taken in February, 1967.
Project Blue Book – USAF photo

NOTE:
There is a fifteen year time span between the 1967 Kreitz UFO photo above and the 1952 Adamski UFO photo below. The appearance of the bottom side of both UFOs is identical.

66

Fawcett
Publications, Inc.

2/2
67 WEST 44th STREET · NEW YORK, NEW YORK 10036 · PHONE 661-4000

January 12, 1968

Mr. Bill Clendenon
P. O. Box 49
Port Angeles, Washington 98362

Dear Mr. Clendenon:

Thank you for your letter of December 20 concerning "The New
Report on Flying Saucers."

Regarding your question of the cover photo, by Kurt Kreitz of
Lancaster, Pennsylvania, I'm sorry I am unable to give you all
the details you wanted. I do know that the photos were in black-
and-white and that there were two snapshots taken.

For the full particulars on the photo, I suggest you write:

> Mrs. Marilyn Stancombe
> Aerial Phenomenon Branch
> Foreign Technology Division
> Wright-Patterson AFB, Ohio 45433

Thank you again for your interest and good luck in your quest.

Sincerely,

Frank Bowers
Editor
Fawcett Books

DEPARTMENT OF THE AIR FORCE
WASHINGTON 20330

OFFICE OF THE SECRETARY

FEB 2 6 1968

Dear Mr. Clendenon:

Your letter of January 31, 1968, addressed to Colonel Mims of the Air Force Congressional Inquiry Division, Office of Legislative Liaison, has been referred to this office for reply.

The photograph mentioned in Colonel Mims' letter is the one appearing in the lower right corner of page 43 of the Fawcett publication, The New Report on Flying Saucers, Issue No. 2.

Sincerely,

DAVID L. STILES
Lt Colonel, USAF
Chief, Civil Branch
Community Relations Division
Office of Information

Mr. Bill Clendenon
P. O. Box 49
Port Angeles, Washington   98362

DEPARTMENT OF THE AIR FORCE
WASHINGTON 20330

OFFICE OF THE SECRETARY

MAR 1 5 1968

Dear Mr. Clendenon:

Your letter of February 2, 1968, addressed to Mrs. Stancombe of the Foreign Technology Division, Wright-Patterson AFB, Ohio, has been referred to this office for reply.

Inasmuch as the picture appearing in True magazine to which you referred was not an Air Force photo, we suggest that you contact the publishers of the magazine for Mr. Kreitz' address. We have no record of it in this office.

Sincerely,

DAVID L. STILES
Lt Colonel, USAF
Chief, Civil Branch
Community Relations Division
Office of Information

Mr. Bill Clendenon
P. O. Box 49
Port Angeles, Washington   98362

CHAPTER THIRTEEN: Unsolved Mysteries

While living in Huntsville, Alabama, I attended a UFO lecture given by Dr. J. Allen Hynek on March 24 1978. Dr. Hynek was billed as our planet's leading expert on the UFO phenomenon. I had corresponded by letter and talked with him by phone a number of times but didn't meet him until this lecture.

During his lecture Dr. Hynek told how he was paid $1,000 for the use of the phrase he coined, *Close Encounters of the Third Kind,* which was used as a movie title. While Dr. Hynek spent time during his UFO lecture down playing the George Adamski UFO case in general, when it came time for the question and answer period Hynek would not answer the questions I asked him in public about his personal investigations of the Adamski UFO case. The reason he gave was the Adamski case was a waste of time to discuss. I thought this was a convenient statement he used to avoid the issue.

Later, I was able to meet with him when we both went to the same restaurant for dinner. He told me that he had spent little time with Adamski and no attempt had been made in using hypnosis in interviewing George Adamski.

Dr. Hynek admitted that his public "chicken feeder" remarks to the news media in Michigan on March 25, 1966, pertaining to the Adamski UFO Scout Ship photo was not based on scientific analysis. I offered again to help him in researching Adamski's UFO case since I had personal experiences but my offer was ignored.

The following two letters from Hynek illustrate how my help and expertise was ignored for reasons I am still not able to understand. The letters are self-explanatory.

DEARBORN OBSERVATORY
NORTHWESTERN UNIVERSITY
EVANSTON, ILLINOIS 60201
20 October 1966

Mr. William D. Clendenon
P. O. Box 49
Port Angelus, Wash. 98362

Dear Mr. Clendenon:

This is answer to your letter of September 21, owing to the presure of many present duties, I have not been able to answer previously.

Thank you for sending the clipping. I am glad to hear that my efforts are being appreciated on the Northwest coast. Perhaps you have seen the October 10th issue of NEWSWEEK (page 70) and very shortly SCIENCE will publish my letter. Also, the Air Force has now announced that the University of Colorado has been chosen to carry out the scientific investigation that I have long recommended.

Coming now directly to your letter: I appreciate your offer of help and the very first time I know of any tangible way in which I can accept this help I will be happy to let you know. Concerning your propulsion system, I am perhaps a very poor person to submit any drawings to since I am decidedly not an engineer and really have very little understanding of mechanical things. I strongly recommend that you find some engineer whom you can trust, so that your patent rights, etc., will be respected, and go over with him in detail the propulsion system you have in mind. Perhaps, also, a chemist would be a good person to consult jointly with the engineer. I am very much afraid that an astronomer is not the person you want.

I am interested, however, in whatever personal sightings you may have made of UFO's. I am enclosing the standard form we use at the Air Force, but I will keep your reply for my own files and not submit it to the Air Force if you would not wish me to do so. I would appreciate very much if you would fill out the enclosed form so that I might add it to my own scientific files which, of course, would be available to the University of Colorado.

As far as Adamsky is concerned, I felt as a result of a personal interview with him, that he was a bit of a faker. At least, he conducted himself in a manner which led me to believe so. It would be profitless, however, to get into such a discussion since I am sure our time can be better spent in other ways. Again let me say that I appreciate your offer of help in the UFO problem and I ask your immediate help in completing the enclosed questionnaire.

Sincerely yours,

J. Allen Hynek

JAH:dz

70

DEARBORN OBSERVATORY
NORTHWESTERN UNIVERSITY
EVANSTON, ILLINOIS 60201

7 November 1966

Mr. William D. Clendenon, Jr.
P. O. Box 49
Port Angeles, Washington 98362

Dear Mr. Clendenon:

I am sorry that I am so very busy at present and that it has taken this
long to answer your letters, but that unfortunately is the way it is. I
get so much mail now that it is impossible for me to answer all of it. My
secretary simply cannot handle it all.

I'm afraid you have really come to the wrong man on the question of propulsion
systems. I am an astronomer and not an engineer and I am more theoretically
than mechanically inclined. From what I can gather from your letter, however,
it does appear that you have given it a great deal of thought and have encountered many obstacles. I can only sympathize with you on that score, but
do not know how to solve it. I have come in for my share of controversey
and misunderstandings also.

I think you are right in that you will not obtain any help from the government.
They receive so many ideas from highly technically trained people that they
in general will not undertake to listen to a layman, because very often the
layman's ideas are found to have some fatal flaw some place. My only advice
would be to find a physicist or engineer near you, obtain his confidence and
friendship and have him discuss your system with you first hand. If he is a
good physicist or engineer, he should be able to tell you in a matter of half
an hour whether your system has any possibilities.

I also forgot to mention that the standard way of proceeding to protect one's
ideas is to obtain a patent on them. This involves some expense, to be sure,
but it is a standard way of doing things. If I had an idea for a propulsion
system for which I wished to retain full credit, I would first see a good
patent lawyer and ask what steps I need to be taken. I believe one has to
have working drawings and possibly a working model. This again requires some
cash outlay and it has been the history of inventors all through the past
centuries to obtain financial help from sponsors, friends and wherever they
can.

Mr. Clendenon   Page 2   7 November 1966

I'm sorry to be of so little seeming help to you, but I am over-committed
at present and simply could not take the time, even If I were competent to
do so to pass judgement on your ideas. In the meantime, I wish you the best
of luck, because luck is what it will take to find the financial aid you are
seeking.

                         Sincerely yours,

                         J. Allen Hynek

JAH:lp

There was a time when I was on the phone with a lady employed at Hynek's headquarters at Northwestern University when we were deliberately cut off right after the woman conveyed a piece of UFO research information to me. When I tried to call her back, I was told that she was unavailable. Once again it appears the public has been led away from the truth.

In July, 1979, I spoke about the George Adamski UFO case on the John Hickons call in radio talk show, station WVOV-AM radio, Huntsville, Alabama. After the talk show, I was invited by John Hickons to appear on his show the following Saturday to do an hour special about the Adamski UFO story.

A few days later, July 16, 1979, I was told not to show up for the Hickons radio talk show as the John Hickons Show had been permanently canceled and John, himself, had been fired. I was never able to make contact with Hickons again.

Was this another case of coincidence?

For a number of years, I had heard of William Spaulding, founder of Ground Saucer Watch, and his computer photograph analysis system used to evaluate UFO pictures taken by civilians including the ones by George Adamski. After corresponding with Spaulding and talking with him on the phone a number of times, I attended his UFO lecture on April 20, 1979, in Huntsville, Alabama High School. Spaulding said during his lecture that Adamski's UFO photographs did not pass his computer analysis investigation. During the question and answer period, he admitted he did the electronic photograph analysis of Adamski's UFO pictures from magazine copies of them. I told Spaulding that this type of analysis was not valid under those circumstances and gave him my reasons.

The originals should be used and one must allow for any distortions caused by ionized air, heat, and movement of the UFO. The audience agreed with me. At this time, I made a public offer of lending him a first generation print from the original Adamski UFO Scout negative so he could conduct his computerized electronic analysis on it. In return, he guaranteed to return the UFO photograph to me along with the written results of his electronic analysis. All of these things were done, however, I have never been given the written results of Spaulding's electronic analysis. I have the agreement in writing dated July 24, 1979 that states the terms of our public agreement. It reads as follows:

W.D. Clendenon
505½ Fountain Row
Huntsville, AL 35801

ADAMSKI UFO PHOTO
ELECTRONICALLY
ANALYZED BY G.S.W.
7/24/79

The enclosed photo was electronically duplicated and is being returned. an analysis, utilizing computer techniques will take place within a few weeks. GSW will forward the results.

GSW
7/24/9

In 1980 my employment in the Huntsville area ended so I took a job on the off-shore oil rigs in the Gulf of Mexico. During 1980, I learned that Alice K. Wells had died.

Again, I was contacted by Wendelle C. Stevens and an opportunity presented itself for producing a working model of the Adamski UFO. In 1981, Stevens was the catalyst which put me in contact with Dennis Edmondson, an aerospace employee in Washington state. He had degrees in science related fields and the funds and facilities to produce a working model.

I discussed my project with Stevens and Edmondson by phone and mail and agreed to send Edmondson enough research material to get him started in the right direction. As usual, I held back material as well as taking steps to protect my work in every way possible.

I sent my research to Edmondson's office by certified mail. He called sometime later and warned me not to send any more research material to his office by certified mail. He stated that he did not want to sign any more receipts and that one morning when he arrived at work, people were going through the material I had sent him. He informed me that he was shown legal papers by the people searching his office. The company where he was employed was working on government contracts involving national security and, therefore, he could do nothing about the search.

He told me that he had invested a bit of money and materials toward producing the working model and that tests were producing positive results. We agreed that all future contact would be at our homes. This was the last contact I was ever to have with Edmondson.

After a period of time, I called Edmondson and was told that he was no longer employed by Western Gear. They would not furnish me any further information. When I tried to contact his home, the phone had been disconnected. All mail I sent to him was either unanswered or returned to me. Wendelle Stevens had also been informed of these events. It seemed Edmondson just disappeared. Coincidence???

In 1981, I was working on the off-shore oil rigs in the Gulf of Mexico. I moved from one rig to another by helicopters or crew boats. Sometimes I was off shore for six weeks, working twelve hour shifts. During my time off, I slept, read, watched television, and talked with geologists and engineers, or fished.

Although an off-shore rig is a good place to watch for UFOs, I never saw one in all that time.

There was one mysterious event which took place on one of the off-shore rigs which puzzles me to this day. Being a relative newcomer to the life on the rigs, I had not discussed my hobbies or interests with other hands on the rig. I spend much time alone just thinking about the UFO phenomenon. It had been a few years since I had seen a UFO and Mrs. Wells' death left me wondering if my UFO involvement was at an end. While Alice K. Wells was alive and during my contact with her, concrete things always seemed to be happening. Now, it appeared that my involvement was in a lull.

As an experiment, I got a copy of a Marine Chart showing our oil rig's exact position in the Gulf of Mexico and included the following information: the rig's name, number, and the name of the firm that owned and worked it, plus the crew change schedules. I included the time I had left on the rig plus my shift hours, the number of my room, and my bunk number. I sent this package to an old address given to me by Mrs. Wells years ago as a place to send material with no questions or comments.

I sat back and waited, saying nothing to anyone on the rig. I thought perhaps the contact would result in a UFO flying over our oil rig. I waited two weeks, as this was the period of time it generally took for results when I was sending material to Adamski.

When nothing happened after two weeks, I resigned myself to the idea that this was a silly idea that wasn't going to produce any results.

Our rig, a large one with a big crew, had an off-shore position and a large helicopter landing deck which made our rig just right to act at an off-shore transfer point for helicopter and supply boats. On certain days regularly scheduled and unscheduled helicopter flights and boat traffic was heavy. There were strange personnel coming and going, men and women, all races, ages, and descriptions so nobody paid attention to the amount of traffic.

My sleeping quarters had four men occupying it, two on duty and two off at any given time. Our quarters were also used to store galley and food supplies as it was the first room next to the galley. The door was open about 50% of the time. People were coming and going all the time.

On that particular day, I went on duty at 5:00 a.m., was in and out of our quarters several times during that morning helping to store fresh food supplies. During the first three trips, I noticed nothing unusual. However, just before the noon in-shore helicopter flight, I made one more trip to my quarters with more food stores. There on my bed lay a plastic, dime-store variety figurine of King Tut's head. It was gold in color, 3 inches by 3 inches in size, light in weight.

Without drawing too much attention, I asked all persons in my room area at that time if they knew who it belonged to and got negative responses. Several had seen it on my bunk but thought it was mine. I spent the remaining time I had on that off-shore rig trying to find out how and why King Tut's likeness came to be on my bunk.

I had not discussed with anyone on that oil rig about my interest in UFOs and ancient Egypt. I had done a great deal of research in regard to ancient Egypt and the Pyramids as well as the symbolic meanings of the Egyptian asp.

To this day, the mysterious incident remains unsolved.

Mask of Tutankhamun shown approximately actual size.

Book 2
UFOs are Real
Mercury's Secrets / UFO Propulsion
Earth's Interior

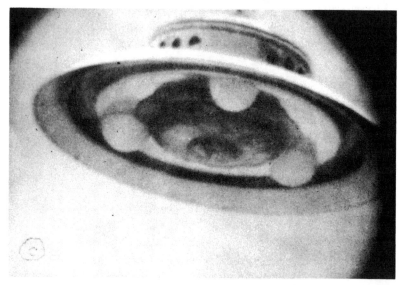

A combination of Clendenon's UFO encounters in the field, plus research of available recorded UFO information from the ancient past to the present has resulted in an analysis leading him to conclude that the UFO in the above photograph is in actuality a Vimana or Space Shuttle Scout / Aero-Space-Plane (A.S.P.). This A.S.P. is not interplanetary. The above machine is a V.T.O.L. (vertical take off and landing) Aero-Space-Plane that appears to be propelled by a binary (two stage) gas turbine (electromagnetic). The hot parts of the turbo pump system are turned inside out to withstand very high operating temperatures. Note mirrored, black and white surfaces of the aircraft's turbo-pump propulsion system. Overall the craft appeared white in color.

Book 2
TABLE OF CONTENTS

THE MYTHICAL GOD, MERCURY

Mercury, the ancient mythical Roman and Greek god, often referred to as the Messenger of the Gods (Mercury, Latin; Hermes, Greek) is described as a male-helmeted figure with winged feet carrying a caduceus, a winged staff with two deadly serpents (ASPs) coiled around it.

Mercury was said to be born of Zeus and the nymph, Maia, in a secret and mysterious cave.

Though often referred to as a young God in myth, in reality, he is one of the oldest. Mercury, as the legends, go, was a hell-raiser and never cared much for higher moral or philosophical developments of religion. The legends state that Mercury was the god of swiftness or flight. It is said if the gods wanted to communicate, carry on commerce, to move things swiftly from one place to another over a long distance safely, they made use of Mercury to accomplish this.

The legends go on to say that because of his nature (unchained energy) Mercury was given a subordinate roll, in the myths of the Gods, that being a messenger or herald of the great gods, an official bearer of important tidings; whose duty it was to announce publicly messages or news from sovereign powers and, on occasion, plead the cause of those who sent him.

Mercury: Messenger of the Gods.

Note: The Caduceus.

THE CADUCEUS, ANCIENT SYMBOL OF POWERED FLIGHT

Mercury, the Messenger of the Gods, carried with him his magic wand or caduceus, the winged staff, with which he could perform many wondrous feats. In one form or another, the ancient symbol of the caduceus appeared throughout the world in ancient history but its true origins are lost in time.

Ancient history of India tells of the *Kundalini,* the fiery serpent force coiled in the center of the Earth, the symbol which is the caduceus staff of the god Hermes, or Mercury, a rod entwined by two serpents and topped with a winged sphere.

In Europe during the Middle Ages, the caduceus of Mercury shomehow became associated with Alchemy and Survived as a chemical symbol and is used by the medical profession universally. It should be noted that a more appropriate medical symbol, the staff of Aesculapius, the Greek god of healing, is a staff with one serpent or snake coiled about it, is being reinstated for symbolic use such as logos by the modern medical profession.

Research indicates that the modern medical profession adopted Mercury's winged staff because the wings represent swift movement or the ability to get about quickly while the two serpents represented chemical or medical meanings.

Some present day aircraft manufacturers use the symbol of the god Mercury's staff, the caduceus, as a logo on their letterheads. Westland Aircraft of England is one, as is the Society of British Aerospace Companies.

The following is a reproduction of the letterhead of Westland Aircraft Limited.

WESTLAND AIRCRAFT LIMITED

YEOVIL
ENGLAND

OUR REF A/LHH/JME/1,307A/"C"

TELEPHONE
YEOVIL   5222
EXT No
TELEX   4677

YOUR REF

TELEGRAMS
AIRCRAFT TELEX YEOVIL

GROUP PATENT DEPARTMENT

14th June, 1965.

Mr. William D. Clendenon, Jr.,
P.O.Box 926
Portland,
Oregon 97207
U.S.A.

Dear Sir,

We thank you for your letter of the 1st June, 1965 regarding your proposal for an aircraft propelled by an electro-magnetic motor, utilizing a mercury ballast for directional control etc.,

Obviously we are interested in examining all new proposals but prefer that you apply for patents before submitting them to us.

If therefore you are prepared to send us a copy of your patent application or specification, we will then examine your proposals to see whether they are likely to be of interest to us.

In order to avoid any misunderstanding we make it a rule to examine inventions on one understanding only, namely, that nothing we do with the invention shall be held by the inventor as rendering us liable to him except to the extent to which the invention becomes a subject of a written agreement entered into between us or, in the absence of such an agreement, we infringe a valid patent granted to the inventor.

Yours faithfully

L. H. Hayward
Patents Manager

Note the caduceus (upper left) as part of westland Aircraft's logo.

Figure 1

The Caduceus, Magic Wand of Mercury, Messenger of the Gods, is an ancient symbol of electromagnetic flight and cosmic energy.

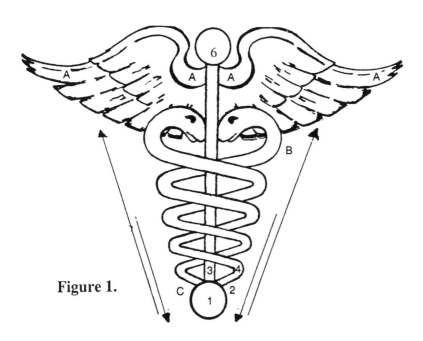

Figure 1.

Key:
A. Air is the flight propellent/propeller/wings
B. Expansion of vortex coils/cooling
C. Compression of vortex coils/heating

1. Liquid metal mercury, the bearer of electromagnetic energy
2. Mercury boiler
3. Antenna/starter/core
4. Closed circuit serpentine (ASP) heat exchanger/condenser coils
6. Poisonous mercury vapor reservoir

If the caduceus of Mercury is a medical symbol, why do present day aircraft manufacturers also use it as a logo? Why is the caduceus sometimes referred to as an ancient symbol of flight? The answers lie not only in the ancient history of Greece but more so in the ancient history of India. The answer is to be found in ancient Sanskrit.

AESCULAPIUS

THE GOD OF MEDICINE AND HIS SYMBOL
The serpent was sacred to Aesculapius, god of medicine, because it was supposed to have healing powers. The god's clublike staff, with a serpent coiled round it, is a symbol of medicine.

MERCURY, THE LIQUID METAL

About 4-1/2 billion years ago, when the planet Earth was formed, the liquid metal mercury (HG), being an element, was included in its make-up. During the passage of time the heavy mercury (with the aid of heat and centrifugal force) worked its way from the Earth's interior outward to the Earth's surface.

It is not known who first discovered the liquid metal mercury, however, it is known that the ancient Chinese, Egyptians, Greeks, Hindus, and Romans knew about this metal. The metal was named for the swift messenger of the gods in Roman mythology.

Mercury is a heavy, silver white metallic element (symbol HG). The symbol for it is derived from the Greek word *Hydrargros* meaning water, silver or liquid gyro. It is liquid at ordinary temperatures and expands and contracts evenly when heated or cooled. Being denser than lead, mercury is heavy and is also known as Quick Silver.

Mercury, the pure liquid metal, has numerous uses. It is written that the ancients referred to it as the metal of a thousand uses. Mercury and mercury compounds have many uses in our world of today; its use is widespread throughout agriculture and industry. As a result, mercury has become widespread in the pollution of Earth's environment. It is interesting to note that mercury pollution of Earth's atmosphere has increased dramatically along with UFO sightings from the late 1940s to the UFO reports of the present time.

The liquid metal mercury, when heated by any means, gives forth a hot vapor that is deadly poisonous in its own right. And, if the liquid metal mercury *is made radioactive,* and heated sufficiently to emit radiation, any leaks in the mercury circuits, such as a mercury vapor turbine may have would, therefore, be a double danger to the crew and maintenance personnel of any vehicle powered by a mercury vapor turbine. Then, too, the liquid metal mercury can be absorbed through the skin of any persons working with it over a period of time.

For shipment and storage mercury is confined in cast iron, wrought-iron, and spun steel bottles or flasks. Sometimes mercury may be confined in ball storage tanks for shipments as the spherical containers has the inherent advantage of being roughly twenty-five percent lighter than other designs and construction if made of two formed spherical heads welded together.

In many electrical devices in industry mercury is imprisoned in glass tubes. And so, the deadly mercury and its vapor is confined or used in closed circuit units in one form or another throughout industry the world over.

However, in spite of the drawbacks, mercury vapor turbines now in use have proved very efficient. Present day mercury vapor turbines use large quantities of mercury but little is required for renewal because of its use in closed circuit systems.

In the industrial world the use of liquid metal mercury is invaluable for use in vacuum pumps, liquid seals, electrical contacts (mercury switches), ultraviolet and fluorescent lamps (mercury vapor is rich in ultraviolet rays) as a cathode in electrolysis. Mercury also remains liquid over a wide range of temperatures. Mercury and its vapor conducts electricity, its vapor also a source of heat for power useage. Mercury amplifies sound waves and doesn't lose timber in quality. Ultrasonics can be used for dispersing a metallic catalyst such as mercury in a reaction vessel or a boiler. High frequency sound waves produce bubbles in the liquid mercury when the frequency of the bubbles grow to match that of the sound waves the bubbles implode, releasing sudden bursts of heat.

Other developments include mercury clutches for electric motors in air conditioning units. But, by far the greatest usage of mercury today is in electrical applications. However, mercury is also used in precision castings, patterns have been made of frozen mercury, a technique used particularly for the casting of complex parts.

Some possible space age uses of mercury is as the propellant in a low thrust ionic engine for long journeys in deep space. Another possible space application is a mercury-filled flywheel for space craft stabilization.

For spacecraft stabilization, and fore and aft propulsion, of the aerospace plane (Adamski's UFO Scout Ship), we might harness the direction-sensing capabilities of the atom by employing three nuclear gyroscopes placed 120 degrees apart on the rotating stabilizer flywheel (25) of the plane. The aerospace plane (ASP) may use the three liquid mercury proton gyroscopes as a means of reference in space flight.

The nuclear mercury proton gyro has several advantages: 1) heavy protons found in mercury atoms are the most stable; 2) it does not require a warm-up period as mechanical gyros do; 3) the gyro using stable mercury protons is not affected by vibrations and shock; 4) the nuclear gyro has no moving parts and can run forever; 5) the atom offers the most stable gyro device in nature and has the additional advantages of saving space and weight. This is particularly valuable on long distance flights where all space and weight must be very carefully calculated and conserved.

DIRECTIONAL GYROS

(1) Three mercury proton gyros or sensor units (29) are mounted on and rotate with the rotating stabilizer flywheel (25) of the aerospace plane. The three sensing cells (29) are rigidly attached 120 degrees apart on the rotating flywheel (25) of the ASP. (2) The three movable coils of the sensing cells (29) are constantly moving in and out or back and forth each in turn as chosen by the computer when the ASP flies on a straight course and (25) is rotating. Signals will be generated by the three mercury proton gyros resistance

to the three coils movements. The signals can then be measured by computer to determine the speed and direction of the ASP. (3) Sensing the up and down motions for hovering of the ASP may be accomplished by the same arrangement of the sensors (29) using the same principles.

ASP MERCURY VAPOR CYCLE

Some advantages of mercury as a binary cycle vapor can be briefly stated as: A) At high temperatures its vapor pressure is moderate; B) The liquid mercury (1) is of sufficient density to be returned to the boiler (2) by gravity, thus eliminating a boiler feed pump; C) Its high density results in moderate spouting velocities throughout all the closed mercury circuits which includes (2), (4), (6), and (29) as employed in this simple binary cycle mercury vapor/gas turbine design; and, D) Mercury exhaust passages can be small, thus the size of the electromagnetic heat exchanging mercury condenser units of (4) and (29) need not be excessive.

NOTE: THERE IS AN OFFICIAL MOVEMENT TO RESTRICT ACCESS TO THE LIQUID METAL MERCURY FOR RESEARCH AND DEVELOPMENT PURPOSES NOT UNDER GOVERNMENT CONTROL.

श्रीमहर्षिभरद्वाजप्रणीत
"यंत्रसर्वस्वा"न्तर्गत

वैमानिकप्रकरणम्

श्री बोधानन्दमुनिविरेण्य प्रणीत
विवृत्या समुल्लसितम्

श्री पण्डिततर्य सुब्बारायशास्त्रिभिः
दैवकृपया समुपलभ्य
हस्तलिखित मातृकारूपेणानुगृहीतम्

उभयभाषाविदुषा
गोमठं रामानुज ज्यौतिषिकेण
यथामति संशोध्यपरिष्कृत्य
आंग्लेय भाषान्तरेणगाह

मैसूर् कार्गेनिवन् मुद्रणालये
लोकोपकागय मुद्राप्य प्रकटीकृतम्

|| ओम् तत् सत् ||

THE VIMAANIKA SHASTRA
An Ancient Treatise on Aeronautics

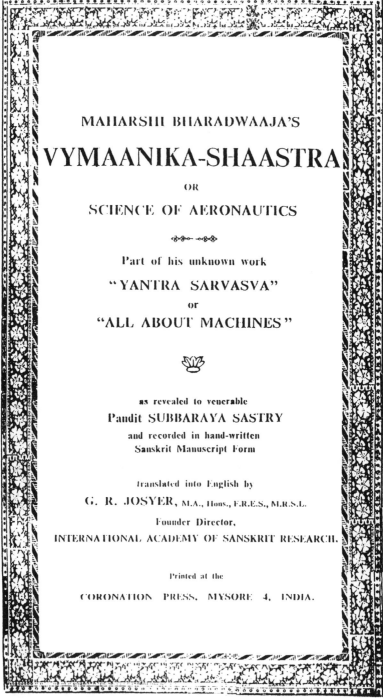

MAHARSHI BHARADWAAJA'S

VYMAANIKA-SHAASTRA

OR

SCIENCE OF AERONAUTICS

Part of his unknown work

"YANTRA SARVASVA"

or

"ALL ABOUT MACHINES"

as revealed to venerable

Pandit SUBBARAYA SASTRY

and recorded in hand-written
Sanskrit Manuscript Form

translated into English by

G. R. JOSYER, M.A., Hons., F.R.E.S., M.R.S.L.

Founder Director,

INTERNATIONAL ACADEMY OF SANSKRIT RESEARCH.

Printed at the

CORONATION PRESS, MYSORE 4, INDIA.

Original Title Page of Josyer's English translation of the
Vymaanika-Shastra, found in 1908 in the Royal Baroda Library.

ANCIENT MANUSCRIPTS

By further correct decoding of the ancient writings in Sanskrit, the classical language of India and Hinduism, along with proper interpretation of the legends of Mercury, Messenger of the Gods, and his magic wand, the caduceus, I discovered a mass of fascinating information about ancient flying machines that might be applied to the advanced aerospace technology of our times.

From the ancient *Vymaanika-Shaastra* (Science of Aeronautics), I gained useful information concerning *Vimanas* or aircraft. Also in the *Samarangana Sutradhara,* it is written that the *Vimanas* (airships) were made of light material with a strong bell-shaped body. Iron, copper, and lead were used in their construction. They could fly great distances and were propelled by air. A hint is then given concerning their propulsion by the statement that they had "fire and mercury at the bottom."

"Strong and durable must the body be made, like a great flying bird of light material. Inside it one must place the mercury engine with its iron heating apparatus beneath. By means of the power latent in the mercury which sets the driving whirlwind in motion, a man sitting inside may travel a great distance in the sky in a most marvelous manner."

"Four strong mercury containers must be built into the interior structure. When these have been heated by controlled fire from iron containers, the *Vimana* develops thunder power through the mercury. And, at once, it becomes like a pearl in the sky."

"Moreover, if this iron engine with properly welded joints be filled with mercury and the fire be conducted to the upper part, it develops power with the roar of a lion."

The Samar says, "By means of these machines, human beings can fly in the air and heavenly beings can come down to Earth."

James Churchward, author of the popular "MU" series, stated that he was shown ancient drawings and instructions of an airship by Hindu priests during his travels in India. The ancient manuscripts described the machinery, engines, and controls to take power from the Earth's atmosphere in a simple inexpensive manner as the aircraft flew along. It was stated that the engine was similar to our present-day jet turbines in that air flows or does work from one chamber into another until the air is finally exhausted. Churchward says, "Air was used as a propellant in something resembling a jet engine."

The key words, here: *air, machinery, engines, turbines, propellant* and *exhausted* sound very familiar in light of the following. Today there is international competition to produce an aerospace plane (ASP) an aircraft which can take off from Earth, fly through the atmosphere, put itself into orbit, and return through the atmosphere to make aerodynamically-controlled landings at an airport.

Today's ASP designs incorporate the idea of carrying machinery in the craft which can liquify the atmospheric oxygen available at low orbital altitudes during flight. This would provide the fuel necessary for orbital maneuvering and reentry. The power plant is known as L.A.C.E. or Liquid Air Cycle Engine.

Over the past 40 years authors have used the above quotations, all or in part, which are from ancient Sanskrit of India and are thousands of years old.

ADVANCED AEROSPACE TECHNOLOGY/ANCIENT MERCURY ENGINES

To avoid any unnecessary confusion, I would like to emphasize the following: Any and all patent rights pertaining to the propulsion system for the Adamski-type UFO Scout Ship legally and morally belong to the people who have designed, manufactured, and are flying the ships. I understood and accepted these facts from the beginning. However, in almost every case in dealing with governments, aircraft manufacturers, institutions of higher learning, I was told to apply for patents on the propulsion system findings my UFO research has gleaned for me and then to submit my patents on the A.S.P. propulsion system to the aforementioned for further consideration.

In some cases a working model was required. Patent lawyers informed me that there could be a need for a number of patents instead of just one. Thousands or more dollars would be required for patent applications. If I succeeded in being granted patents for the A.S.P. propulsion system, the authorities were apt to step in and control things after that. In addition, a number of aircraft manufacturers would not deal with individuals pertaining to unsolicited inventions unless the individual was employed by that company.

Therefore, I took steps when making any of my research work public to protect my work in the only ways that were available to me. I have always informed individuals, publishers, and lecture audiences, that I had in some manner protected my research effort while making them public.

If ever a working model of the Adamski-type UFO Scout Ship is produced anywhere in the known world, those who produce it should expect a strong response, from the mysterious people who now operate this type of aircraft.

As NASA was involved in research programs pertaining to the use of mercury and/or nuclear energy for use in future aerospace propulsion systems, I submitted some of my research findings to NASA for evaluation. In my efforts to generate funds and interest in the ASP propulsion concept, I also contacted U.S. Senator Warren G. Magnuson of Washington state asking for his assistance in obtaining a government grant to aid me in my research efforts, while at the same time asking him to obtain an official statement from NASA as to whether or not the ASP propulsion concept was feasible. In answer to these requests, I received the following letters.

NATIONAL AERONAUTICS AND SPACE ADMINISTRATION
WASHINGTON, D.C. 20546

OFFICE OF THE ADMINISTRATOR
C:Evvb:A14350f

MAY = 1967

Honorable Warren G. Magnuson
United States Senate
Washington, D. C. 20510

Dear Senator Magnuson:

We wish to reply to your recent letters on behalf of Mr. William D. Clendenon, Jr., of Port Angeles, Washington, who wrote you concerning his "ASP propulsion concept," a proposed atomic powered aero-space plane. He indicated that he had submitted the concept to the Inventions and Contributions Board of the National Aeronautics and Space Administration some time ago. He has requested an "official" statement that his concept is feasible, to assist him in his development of the concept.

NASA scientists have examined in detail Mr. Clendenon's concept and have concluded that all the technology necessary to its implementation is not now in existence. Although it is possible that his ideas may receive application in the

94

future, we are not in a position to predict this
with any degree of certainty. Mr. Clendenon has
previously been informed of the technical difficulties
and problems which require solution and are pre-
requisite to the particular applications he foresees.

We have also informed Mr. Clendenon that his ideas
do not qualify for monetary award for the reason
that they have not been developed and used to the
benefit of the Government.

We appreciate having had the opportunity to review
Mr. Clendenon's proposal.

                         Sincerely yours,

                         Richard L. Callaghan
                         Assistant Administrator
                            for Legislative Affairs

May 8, 1967

Dear Mr. Clendenon:

     For your information. If you feel there is anything
further that I can do, please feel free to contact me again.
Your drawings are returned with this letter.

     Kind regards.

                         WARREN G. MAGNUSON U.S.S.

WGM:b

As an eye witness I can truthfully state that the UFO in
the Adamski Scout Ship photograph of 1952 is an
intelligently-controlled flying machine. It is also my belief
that this particular UFO photograph, when correctly analyzed, will be the key to help unlock the UFO mystery or at
least the true part George Adamski played in it. Simply, the
results of years of experience and research in the UFO field
points again and again to the following possibilities:

That the UFO in the 1952 Adamski Scout photograph is a
Vimana or aircraft of ancient design (a Model-T perhaps).

This aircraft is in reality a V.T.O.L. aerospace plane propelled by an air breathing turbo-pump propulsion system designed basically for flight within the planet Earth's atmosphere and, therefore, is not interplanetary. As the propulsion system may be multi-fuel in design, it may also be a liquid air cycle engine (L.A.C.E.) that collects and liquefies air for fuel from the Earth's atmosphere as the aircraft passes through it. Note that because of the A.S.P.'s possible multi-fuel design, it may also use Jet Plane fuel, when desired.

The basic turbo-pump engine has four main sections: compressor, combustion, or heating chambers, turbo-pump and exhaust. Burning gases are exhausted through the turbo-pump wheel to generate power to turn the electric generator:

(1) Propellant tanks will be filled with liquid air (obtained directly from the atmosphere by on-board reduction equipment).

(2) Liquid air may be injected into expansion chambers and heated by the metal working-fluid mercury confined in a boiler coupled to a heat exchanger.

(3) The super heated M.H.D. plasma (or air) will expand through propellant cooled nozzles.

(4) The ship may recharge its propellant tanks with liquid air and condensate water collected directly from the upper atmosphere by the on-board reducing plant.

In researching the mercury engine in the ancient Sanskrit *Samarangana Sutradhara,* the text pointed in the right direction. However, the problem was to bring together the Sanskrit text with my research findings. To fill in the gaps was the challenge. As an example, to explain more clearly, I have inserted technical points to some of the original text as follows:

The Sanskrit *Samarangana Sutradhara* says, "Inside one must place the mercury-engine with its iron heating apparatus beneath." -- *Inside the circular air frame, place the mercury-engine with its electric/ultrasonic mercury boiler at the bottom center.*

"By means of the power latent in the mercury which sets the driving whirlwind in motion a man sitting inside may travel a great distance in the sky in a most marvelous manner." -- *The unchained heat energy from the hot mercury vapor sets the air pump/turbine in motion.*

"Four strong mercury containers must be built into the interior structure. When these have been heated by controlled fire from iron containers, the *Vimana* develops thunderpower through the mercury. And at once it becomes like a pearl in the sky." -- *One mercury boiler and three mercury flux valve sensor units must be installed in the propulsion system within the center of the circular air frame. When these mercury containers have been heated by electrically-controlled fire (hot mercury vapor) from the containers, the aircraft develops ultrasonic power through the mercury. And, at once, the ionized recirculating air flow becomes like a pearl in the sky (M.H.D.)*

NOTE: Ann Grevler claims to have been on flights in an Adamski-type UFO. Tim Good (UFO Researcher) came across fascinating information on propulsion from Ann's story.

"... The general idea of their propulsion is that cosmic power (electricity?) is drawn out of the surrounding air, through the top of the central column ... This power is then irradiated via a pump at the bottom of the central pillar, over powdered quartz of a kind, which is spread over the largest possible field within the ship. The result is ionized air ... this is pumped through the three hollow rings around the outside base of the cabin structure as well as circulating through the three balls underneath — these latter being used for motive power and direction and not used primarily as landing gear ..."

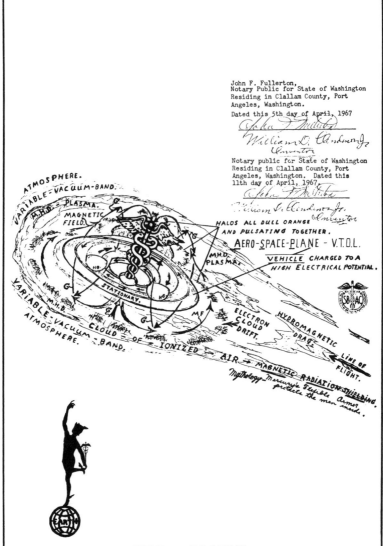

UFO and M.H.D.

NOTE:
SIMPLY PUT . . . THIS TYPE OF UFO OR AERO-SPACE-PLANE APPEARS TO BE A GYRO MOTOR. M.H.D. PLASMA ACTS AS AN OUTSIDE ROTOR, THE CADUCEUS COIL OF MERCURY IS EMPLOYED INSIDE AS THE FIELD COIL OR STATOR.

THE UFO AND M.H.D. PLASMAS

The UFO Scout Ship is often seen as a ball of light. This ball of light is an M.H.D. plasma that surrounds the UFO because hot, continuously recirculating air flows through the ship's gas turbine and is ionized (electrically conducting). At times a shimmering mirage-like effect caused by heat, accompanied by pulsations of the ball of light makes the UFO appear to be alive and breathing. This has, at times, caused people to mistake the UFO for a living thing. For some of the above reasons, the ship may seem to suddenly disappear from view, though it is actually still there and not de-materialized, as some people have claimed. The ionized bubble of air surrounding the UFO may be controlled by a computerized rheostat so that the ionization of the air may shift through every color of the spectrum obscuring the aircraft from view.

The M.H.D. plasma or recirculating cloud of ionized air surrounding the UFO Scout Ship may reduce the abruptness of shock waves when the aircraft is moving through the Earth's atmosphere.

Cancellation or reduction of sonic boom and engine noises (silent flight) may be brought about as the pulsating M.H.D. plasma results in a variable vacuum bubble surrounding the machine, thus isolating the UFO and its M.H.D. plasma from friction with the Earth's atmosphere at times.

The M.H.D. plasma also acts as magnetic radiation shielding for the UFO or ASP while operating at the higher altitudes.

When the UFO or ASP decelerates in a straight line of flight, a point of light develops from the center of the aircraft outward, resulting in a surrounding M.H.D. plasma bubble or ball of light engulfing the aircraft. As the A.S.P. accelerates, the surrounding ball of light or plasma fades out.

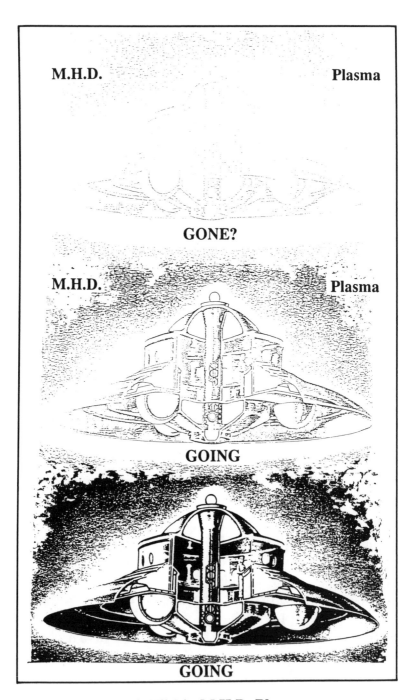

UFO Within M.H.D. Plasma

An atmospheric side effect UFOs display is to discharge a cloud-like vapor while hovering, giving the ability to hide in or appear as a cloud. While hovering at night, the UFO sometimes gives the appearance of a bright star in the sky. Thus, an example of the UFO's excellent built-in camouflage mechanism.

All of these technical clues again point strongly to my conclusion that the Adamski-type UFO Scout Ship is propelled by some sort of air-breathing turbine.

ELECTROMAGNETIC LIFT/REPULSION

1. A complex machine or aerospace plane having a circular air frame design (32), may consist partly of a powerful electromagnetic (4) through which is passed a rapidly pulsating direct current.

2. Electromagnetic field coil (4) (closed circuit heat exchanger/condenser coil circuit containing the liquid metal mercury (1) and/or its hot vapor) is placed with its core axis (3) vertical.

3. A ring conductor (25) (directional gyro-armature) is placed around the field coil (heat exchanger) windings (4) so that the core (3) of the vertical heat exchanger coils (4) protrudes through the center of the ring conductor (25) (directional gyro-armature).

4. When the electromagnet (4) (heat exchanger coils) is energized, the ring conductor (25) (armature) is instantly shot into the air (taking the A.S.P. as a complete unit along with it.)

5. If the current is controlled by a computerized resistance (19) (rheostat), the armature ring (25) A.S.P. can be made to float or hover in the Earth's atmosphere.

UFO/Vimana/Aerospace Plane/A.S.P.

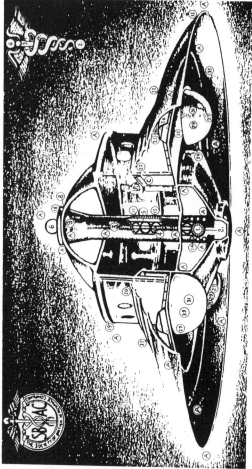

KEY:

A. Ionized-Recirculating Air-Flow-
   M.H.D.-Plasma
C. Computer-Control-Console
P. Pilot

1. Liquid Metal Mercury (H.G.)
2. Mercury Boiler
3. Antenna/Starter/Core
4. Electromagnetic Heat
   Exchanger/Condenser
   (Closed Circuit)
5. Inner Splined Shaft
6. Mercury Vapor Reservoir
8. Air Intake
9. Internal Coanda Nozzle
11. Compression Area
14. Exhaust Ports
15. Outer Splined Shaft
18. Coanda-Shroud
19. Rheostat-Coil
20. Electrical Condensor/Capacitor
    (Pressurized Crew Cabin)
22. Liquid Air Cycle
    Reduction Machinery

23. Fuel Tanks
24. Air Pump (Turbo-Pump) (Rotates
    counter clockwise)
25. Stabilizer Flywheel/Directional
    Gyro/Armature (Rotates clockwise)
26. Cooling-Fan (Rotates counter clock-
    wise)

27. Fixed Sphere (Housing)
29. Flux Valve (Sensor Unit)
31. Mercury Switch (Closed Circuit)
32. Air Frame (Circular)
33. Outer Flange
35. Plenum Chamber
37. Expansion Joints

6. The electromagnet (4) (heat exchanger) hums and the armature ring (25) (or torus) becomes quite hot. In fact, if the electrical current is high enough, the ring (25) will glow dull red or rust orange with heat.

7. The phenomenon (outward sign of a working law of nature) is brought about by an induced current effect identical with an ordinary transformer.

8. As the repulsion between the electromagnet (4) (heat exchanger) and the ring conductor (25) (armature) is mutual, one can imagine the aerospace plane (A.S.P.) being effected and responding to the repulsion phenomenon as a complete unit.

9. Lift or repulsion is generated because of close proximity of the field magnet (4) (heat exchanger) to the ring conductor (25) (armature).

NOTE: *Lift would always be vertically opposed to the gravitational pull of the planet Earth, but repulsion can be employed to cause fore and aft propulsion.*

FORE AND AFT PROPULSION FLIGHT CONTROL

Part I

1. An Aerospace Plane with a circular air frame design (32) and its Liquid Air Cycle Engine (L.A.C.E.) a form of gas turbine mounted vertically, would employ a flywheel (25) containing mercury for stabilization, guidance, and control.

2. Three multi-function inertia sensors (flux valves) (29) containing mercury switches (31) are housed within three fixed globular containers (27) and placed 120 degrees apart. These fixed globular housings (27) are mounted on the underside of the outer rim of the stabilizer flywheel (directional gyro/armature) (25). The three multi-function

103

ADAMSKI'S U.F.O. SCOUT SHIP IS A PRODUCT OF ADVANCED AEROSPACE TECHNOLOGY

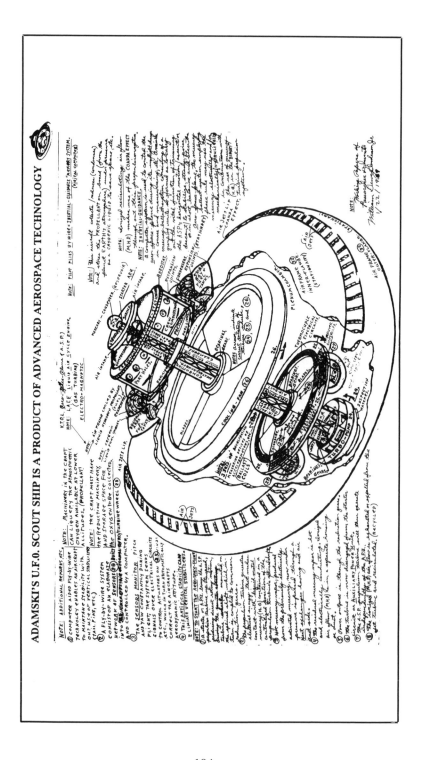

sensor units (29) (electromagnetic ballast coils) when mounted on the directional gyro (25) incorporates in a single device the properties associated with a gyro and accelerometers that can be used to control the attitude of the A.S.P. about three axis as well as cause fore and aft propulsion of the aircraft.

Part II

1. Take a wobbling gyro-like aircraft (flying disk) with a gas turbine propulsion system that is controlled by a computer.

2. Employ a rotating directional gyro (25) as part of the turbine with three electromagnetic mercury sensor units (29) placed 120 degrees on the flywheel (25).

3. By computer control, insert (by centrifugal force) and withdraw (by electromagnetic repulsion) the ballast coils (29) one at a time while placed 120 degrees apart on the rotating flywheels (25) outer rim, the flying disk aircraft will move forward a little on its chosen line of flight.

4. Using the computer, activate the sensor units (29) (in a piston-like action) one after the other, 1, 2, 3, or bang, bang, bang; one right behind the other in the same chosen area or angle of the rotating flywheel (25).

5. Then the flying disk will move forward a little on its flight path each time the sensor units (29) are fired or energized in the same selected area or angle of the rotating flywheel (25).

6. Thus, you will have a continuous propulsion in a straight line of flight, but with a skipping-like forward motion of flight, like a flat stone skipping across the smooth surface of a pond, that will soon take the aircraft over the horizon.

7. The computer controls the angle of attack or attitude of

the flying disk skipping line of flight as the pilot desires.

8. While in all modes of operation, the gyro-like aircraft may, by computer, constantly monitor and control all mechanical forces built up during the propulsion system's operational modes. For example, by activating all three sensor units (29) at the same time along with proper computerized adjustments of the propulsion turbines rpms, the gyro-like aircraft can be made to hover, rise straight up, or descend in a level position.

9. This type of Aero-Space-Plane achieves lift by several means. Electromagnetic repulsion, electrically conducting hot recirculating air flows (M.H.D. plasma) over curved surfaces, the coanda effect and an air cushion in the plenum chamber. (Ground Effect)

Plate 4 Electromagnetic Repulsion of a Conductor

Leonard G. Cramp

The foregoing points in the right direction. The author, however, has taken steps to protect himself and his work, while performing a worldwide public service. The author is willing to assist in any serious R & D program to try and duplicate (in model format at least) the propulsion system of the ADAMSKI type UFO scout ship.
NOTE: *It is my hope that knowledge gained from such a professionally conducted effort would be made available to the public worldwide.*

W. D. Clendenon

THE ORIGINS OF UFOs

What are the origins of the Flying Saucers or UFOs?

After 40 years of research, I can only speculate on the answer; it remains as much of a mystery to me as it is to everyone else. There are many theories but the origins of the UFOs remain a mystery. However, needing a starting point for speculation, I have chosen three areas which are:

EARTH, as it is known in general.

EARTH'S INTERIOR, generally unknown.

OTHER PLANETS, yet unexplored.

EARTH

Given the advanced flying machines that UFOs seem to be, the chances of secret UFO bases having been established on Earth at one time or another by human beings of unknown origins cannot be ruled out. There are still areas of the Earth's surface and sub-surfaces which have not been fully explored or even properly mapped. Unexplored areas may still be found in the polar regions, Himalaya Mountain range, the far North American Continent, the extreme tip of South America, the Great Mid-Eastern Deserts, and the tropical rain forests of the Amazon Basin, New Guinea, and Borneo. UFO bases could be hidden in any of these places.

It just isn't likely that any known Earth government could be responsible for the existence and the operation of these mysterious, intelligently-controlled UFOs because it would be too difficult to keep secret. The problems of manufacturing, testing, logistics, and cost would have to be hidden.

Then international laws would have to be considered regarding UFO over-flights. Considering the numerous designs, types, and sizes of UFO, which have been reported over the years, Earth governments just couldn't claim ownership. The secret of the United States nuclear bombs and the American stealth aircrafts existence could not be kept secret for more than a few years. The 40-year-old UFO mystery of our time is still just that, a mystery.

It is an international mystery which could be solved with cooperation, open-mindedness, and a true organized effort to find the answer.

EARTH'S INTERIOR

An ancient mystery which still confronts modern-day scientists is the internal detailed structure of the Earth. Our scientists have not yet discovered how our Earth functions as a natural working dynamo which produces heat, electricity, water, air, and life. When scientists do discover all the workings of the Earth, mankind will have the answer to the above mystery. Although there are a number of theories, scientists are cautious in making their statements public. I have stated before, the general public would be the last to know if the evidence would surface that the Earth is *hollow.*

Searching the ancient Earth's legends more closely may give clues to the Earth's true history. Ancient theories applied to modern methods of Earth science explorations may be the answer.

Ancient legends about hidden kingdoms, Shangri Las or forgotten worlds, are found throughout history. The ancient Greeks were very interested in the depths of the Earth; many legends refer to it. In Greek Mythology, legends tell us that to reach the land of the Gods, one must travel far to the north, pass the land shaped like a musical harp or lyre (Greenland?), and Mercury, Messenger of the Gods, must be our guide. For Mercury is the driver of the Sun Chariot (UFO?) that flies as fleet as thought. Mercury is the keeper of the gateway to the underworld, home of the gods. The Christian Bible tells us of a hell world beneath our feet (a cover story?) but could it be a green land with a gateway that opens and closes naturally, possibly an icecap?

The Book of Job 38:8 says, "Who shut up the sea with doors, when it break forth, as if it had issued out of the womb?" And, in Job 38:22-23, we read, "Which I have reserved against the time of trouble, against the day of battle and war? Out of whose womb came the ice and the hoary frost of heaven, who hath gendered it?"

The effectiveness of cloud cover or fog in hiding a secret is well-known. A snow-capped mountain was discovered in tropical New Guinea several years after World War II. The mountain was not discovered on previous flights because of the constant cloud cover in that area. There are visibility problems because of fog and clouds in the Earth's polar regions which hamper exploration.

Geographers tell us that Greenland is Earth's largest island and if we could break through and lift up the Greenland icecap, we would find a rugged ring of mountains which form a natural bowl. Under that icecap, we are apt to find lots of water. Modern scientists exploring the mysteries of Earth's gravity at a bore hole in Greenland in 1988 found some new and strange information regarding Earth's gravity in the Greenland area.

During space shuttle mission #27, astronauts took detailed photographs of Greenland and these particular photographs remain classified today because they are said to be "of a military nature."

The Arctic Polar Vortex is a region around the North Pole area that becomes isolated from the rest of the Earth's atmosphere during a two-month period each winter. Scientists believe the area suffers ozone depletion during that time due to little mixture of air from other areas during this coldest time of year.

Let us assume for a moment that the interior of Earth has a large cavity within containing gases. Let us further assume that this cavity is an air reservoir. Now compressed air or gases of the Earth's interior air reservoir emerges from north and/or south polar air vents, and is super-cooled by sudden expansion. Ice is formed because the polar air vent areas are super-cooled by the emerging compressed gases from the Earth's inside air reservoir. If the Earth's polar air vents are not open permanently, then the Earth's polar icecaps may act as valves. Heated gases or air under pressure within the Earth's interior air reservoir may cause the polar icecap to melt or open naturally. The emergence and sudden expansion of compressed gases from the Earth's interior air reservoir super-cools and freezes shut (during this coldest time) or closes the Earth's polar air vents, if the polar air vents are not too large. The opening and closing cycles of the Earth's polar air vents is naturally automatic and might well occur in a period of two months.

Ancient legends of India tell of the *Kundalini,* the fiery serpent force in the center of the Earth, and *Fahat,* cosmic electricity within the Earth. The symbol of *Kundalini* is the *Caduceus,* the winged staff of the God Mercury, a rod entwined by two serpents. If the planet Earth was naturally formed by a vortex motion, then the Earth may very well have a cavity within, making the Earth a rotating hollow sphere or gyroscope.

As the Earth is gyroscopic in form and motion, it should obey all the laws of gyroscopes. It should be noted that the liquid metal mercury is contained within the Earth and is a heavy element which is electrically conducting and an emitter of heat. Mercury can be made radioactive when impregnated from some outside source. The chemical symbol for liquid Mercury is HG which is Greek for liquid gyro. The heart of a gyro is the rotor or spin motor driven by electrical energy or jets of gases. The gyro rotor is thickest and heaviest at the rim so that it acts like a flywheel in motion, thus a gyroscopic hollow sphere is a natural to be further formed into a planetary dynamo. The jets of gases that drive the spin motor could be on the inside edge of the rim of the planetary dynamo with the jets of gases exhausted into the interior air reservoir allowing for a recirculating air flow system to be coupled to the Earth's outside atmosphere through the north and south polar air vents. As the

Earth rotates, it sometimes speeds up and slows down, wobbling, expanding and contracting, and according to scientists, ringing at times like a vast bell. A bell, no matter how vast, must have a cavity within filled with a gas (air) in order for the bell to ring. Sound does not travel in a vacuum.

The dynamo theory of Earth's magnetism is the most promising one at present. If the Earth is a hollow sphere, then the main source of the Earth's magnetism may be in the core of the Earth's rim. A huge natural dynamo inside the Earth converts mechanical energy into electromagnetic energy. The mechanical (hydraulic) energy would be supplied by fluid motion, convection, carrying electric current inside the core of the Earth's rim. Rotation of Earth actively contributes in creating the Earth's electromagnetic field. As the core of the Earth's rim is molten and made of heavy substance (mercury), the Earth acts as a huge gyroscope and its axis of rotation remains constant. The heart of the Earth gyro may be a rotor or spin motor driven by electricity and/or interior blast of gases. The magnetic field of Earth is generated by ordinary electric currents circulating in the Earth's interior.

Such currents must be continuously supplied from some source of energy within the Earth. This takes place in the part of the Earth where there is the least electrical resistance, that is, in the fluid core of the rim. Large slow-moving eddies of molten metal (mercury) swirling in the core of the rim is a thermo-electric effect at the boundary between Earth's mantle or shell and the fluid core of the rim.

This characteristic of the core could result from a combination of heat being generated in the Earth's interior by radioactive elements and the rotation of the Earth. This may account for the Earth's electromagnetic field.

The swirling eddies of molten liquid metal (mercury) in the core of the Earth's rim would act somewhat like electromagnets and create a magnetic field.

A.S.P. AERO-STEAM POWERED PLANETARY ENGINE

If the planet Earth was naturally formed by a vortex motion, then the Earth should have a cavity within, making the Earth a rotating hollow sphere or Gyroscope. As the Earth is gyroscopic in form and motions, etc., it must obey all the Laws of Gyros.

NOTE:
The Liquid Metal Mercury (H.G.) is a heavy element, electrically conducting, and an emitter of heat. Mercury can be made radioactive.

NOTE:
The main source of Earth's magnetism may be within the core of the Earth's rim.

NOTE:
The chemical symbol for liquid mercury is HG, Greek for liquid GYRO. The heart of a gyro is the Rotor or Spin Motor driven by electrical energy or a jet of air. The gyro rotor is thickest and heaviest at the rim so that it acts as a flywheel in motion, thus a gyroscopic hollow sphere is a natural to be further formed into a planetary dynamo.

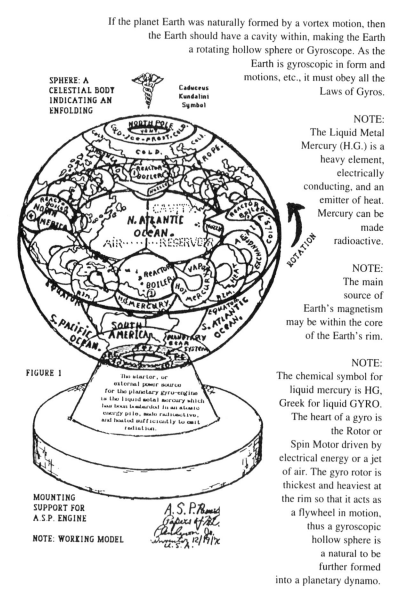

SPHERE: A CELESTIAL BODY INDICATING AN ENFOLDING

Caduceus Kundalini Symbol

FIGURE 1

The starter, or external power source for the planetary gyro-engine is the liquid metal mercury which has been bombarded in an atomic energy pile, made radioactive, and heated sufficiently to emit radiation.

MOUNTING SUPPORT FOR A.S.P. ENGINE

NOTE: WORKING MODEL

112

Plato, a Greek philosopher, spoke of Atlantis around 360 B.C. In general, Atlantis was thought to lie somewhere beyond the Pillars of Hercules, said to be the Straits of Gibraltar, but in the 17th century F. de La Motte LeVayer claimed evidence existed that Atlantis' true position lay in the area of Greenland.

Could it be that under or beyond the Greenland icecap lies the true Green Land, a paradise lost? Is this the true meaning of Job 38:11 when God said, "Hitherto shalt thou come, but no further: and here shall thy proud waves be stayed."?

Alice K. Wells had at times discussed the ancient legends of the Hollow Earth theory with me but strongly discredited the legends as having any basis of fact. I had stopped discussing the subject with her, but she brought it up occasionally. I thought that my research material had impressed her more than I had realized.

May 24,1978

Dear Bill Clendenon.

        The additional info just received. Glad you are finally getting some recognition for your motor.
        You are right about Hynek. The Rodeffer UFO pictures are authentic. George was there at the time the craft appeared and the pictures were taken. I do not know about Allinghams other pictures- but the one of the back of the space man is correct.

        Now about the hollow earth; I know that much speculation, pro and con- and articles and even books have been writen about it --- BUT ** IT IS NOT TRUE, each planet in our solar system has a molten metal chore. This is what keeps each planet in its prescribed orbit around the Sun. Can you imagine what would happen if this were not true ?
        In our analysis of everything in life we have to use common sense. Years ago I red a book about a civilization that liven ih the Earth's interior . This was supposed to be an eye witness report, as it were. No doubt a base that space people used was near the pole and it was thought to be inside the earth. .
        THE HOLLOW EARTH theory has been kicking around Europe for quite some time now. But the theory is not based on facts according to the Brothers that G.A. met.

                    Best regards
                        Sincerely

                        Alice K. Wells

If the planet Earth were indeed a hollow planet and contained in its interior a hidden, highly technical ancient civilization and the Earth's atmospheric air flow was at times recirculated or recycled with the gases of the Earth's interior air reservoir by way of polar air vents, then any outside pollution of the Earth's atmosphere might well affect life forms within the interior air reservoir of the planet. In any event a nuclear war might well destroy the Earth and its inhabitants, inside and out. A nuclear war on the outer regions of Earth would destroy any interior civilization. One needs to ask why the interior civilization chose to make contact with the outer civilization when it would have been safer to remain hidden. Or, maybe they could no longer remain hidden because of the civilization growth on the outer Earth. Therefore, the story of extraterrestrials was manufactured to explain their presence. This would explain how the UFO people that Adamski encountered came and went throughout the Earth's surface civilization with such ease.

The Ancient Hollow Earth theory has been gaining more attention ever since the late Admiral Richard E. Byrd returned from his Polar flights of the late 1940s and early 50s.

The National Examiner, 12, 23, 86, P. 27, referred to Dr. Harley A. Byrd, UFO Researcher. The Examiner pointed out that Dr. Byrd had worked in the offices of the U.S. Air Force's UFO Project Blue Book. The Examiner quoted Dr. Byrd as follows: "My uncle (Admiral Byrd) flew into the center of the Earth and discovered another race, but they wouldn't let him tell about it. We know that there's an inner continent in the Earth."

In the same Examiner article Dr. Keiji Nakamura, a scientist with The Imperial Institute of Astronomy in Tokyo, said his research suggests that an inner continent exists. Dr. Nakamura also stated that The Imperial Institute is planning an expedition to the Artic in 1992 to locate, if possible, an opening to the inner world.

"The experts in defense conceal themselves within the Earth; those skilled in attack move from the Heavens."
SUN TZU.

114

US covers up facts of UFO base under South Pole

STARTLING new evidence from the South Pole proves that inner earth is inhabited by space aliens — and the US intelligence community is covering up the shocking truth!

So reports a frightened veteran of America's Antarctic research program who has just returned from a covert UFO fact-finding mission there. Among his incredible findings:

● Antarctica, the world's highest, coldest, and least explored continent, is a hotbed of UFO sightings and the door to an advanced, possibly hostile civilization that hides in the center of the earth.

● The recent and highly publicized discovery of lost ozone in the atmosphere above the frozen land mass is the direct result of UFO activity there.

● Amundsen-Scott South Pole Station, a little-known multi-million-dollar research base at the bottom of the globe, secretly functions as the world's most sophisticated UFO tracking facility.

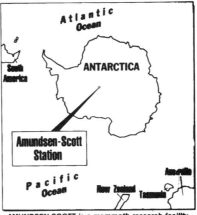

AMUNDSEN-SCOTT is a mammoth research facility.

And the station is quickly sinking into an enormous cavern carved by "inner earthlings" angered by intrusions into their territory.

The investigator, who insisted on anonymity, has been a seasonal employee of the United States Antarctic Research Program for nearly a decade.

Said the investigator: "There were reports that the Amundsen-Scott South Pole Station, our mammoth research facility at the bottom of the world, was sinking into a hole of unknown origin. I was flown there to investigate."

Threading his way through a maze of steel tunnels beneath the base, the investigator and a coworker discovered a small opening in the floor of a steeply sloping shaft.

Explained the researcher: "Hot vapors of what smelled like methane were coming out the opening. We tried looking in with flashlights, but couldn't see a thing.

"So we attached a thermometer to several hundred feet of cord and lowered it in. We never hit bottom, but the thermometer read close to -20 F. That's 40 degrees warmer than the average ice temperature."

Just days later, however, while looking for blueprints of the base in its library, the investigator came across the reports of UFO experts Ray Palmer and Raymond Bernard.

Palmer and Bernard believe most UFO sightings are of craft that originate deep within the earth's hollow interior.

They are piloted by beings of an advanced civilization who have mastered flight both within and beyond our planet. They enter and exit earth at the poles.

Said the investigator: "I'm absolutely certain that Palmer and Bernard are correct. And I strongly suspect that the US intelligence community knows they are, too."

The real reason for the US base at the South Pole is revealed by the nature of the experiments performed there, explains the researcher.

"It's mostly deep drilling into the polar plateau and upper atmospheric probing with rockets, telescopes, balloons, and spy planes — exactly what must be done to track down and contact the inner earthlings," according to the investigator.

He adds: "Antarctic program spokesmen will deny this, of course. They claim there are no covert or classified activities of any sort.

"But the protective shields on South Pole satellite tracking antennas are labeled, 'Langley, Virginia.' Langley is home of the CIA, well known for collecting and suppressing data on UFOs."

— JIM RICHARDS

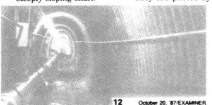

12 October 20, '87/EXAMINER

AN OPENING was discovered in a tunnel floor. Does it lead to an advanced, hidden civilization?

Mountains, valleys found at Earth's core

By LEE SIEGEL   THE SUN HERALD
ASSOCIATED PRESS WRITER

■ SAN FRANCISCO — Crude maps of the **depths** of the Earth show its molten metal core is not a smooth sphere, but instead is roughened by mountains taller than Everest and valleys six times deeper than the Grand Canyon, researchers said Tuesday.

Friction from the sloshing of the liquid across these features may explain why the planet rotates with a slight jerkiness that makes a day five-thousandths of a second longer or shorter than 24 hours every decade, said the scientists from NASA, the California Institute of Technology and a British laboratory.

"There has been no previous evidence for bumps on the core," said Caltech geophysicist Robert Clayton. "They were only speculated. This is direct evidence that they exist."

A Caltech research team's report on the topography of the boundary between Earth's molten nickel-iron core and the surrounding rock mantle, will be presented Wednesday during the American Geophysical Union's fall meeting.

The maps of the core-mantle boundary, located about 2,000 miles beneath Earth's surface, were made using a 5-year-old technique called seismic tomography, in which varying speeds of earthquake waves travel through molten and solid rock are measured.

The technique is similar to a medical CAT scan, which can make images of the inside of the body because X-rays travel in different ways through various types of tissue.

The maps of the core-mantle boundary were made by Clayton, Caltech geophysics graduate student Olafur Gudmundsson and professor Don L. Anderson, who used worldwide records of thousands of earthquakes that occurred between 1971 and 1980 and measured more than 4.5 on the Richter scale.

The maps provide poor resolution, or detail, and failed to map the core-mantle boundary in some places, especially the Southern Hemisphere.

But far beneath the Philippine Sea, the core shows a "low" or valley at least 6 miles deep, more than six times the depth of the Grand Canyon. Beneath the Gulf of Alaska, there is a 6-mile-high mountain on the core — taller than Mount Everest.

Other underground mountains were found under eastern Australia, the central North Atlantic, the northeastern Pacific, Central America and south-central Asia. Valleys exist in the core beneath the southwestern Pacific, the East Indies, Europe and Mexico.

Friction between the Earth's surface features and wind in the atmosphere alters Earth's rotation so that a day really varies in length by one-thousandth of a second over the course of a year.

But until now, scientists haven't been able to explain another "jerkiness" in Earth's rotation, that also makes a day's length vary by one five-thousandth of a second each decade, Clayton said. WEDNESDAY, DEC. 10, 1986

The sloshing effect of the molten core across valleys and mountains at the core-mantle boundary nicely explains that variation, said the NASA-Caltech study presented by Caltech's Brad Hager.

Clayton attributed the core's bumpiness to the tendency of heated materials to rise. Where the core has mountains, the overlying mantle flows upward under great heat and pressure.

THE EARTH—A VAST "BELL"

A STUDY of prolonged resonance effects which result when the earth is struck a hammer blow by a major earthquake, indicates that the earth "rings"—and that, in fact, it is more resonant than the most perfectly cast bell.

The discovery of the earth's "ringing" was reported by Dr. Gordon J. F. MacDonald of the University of California, in Los Angeles, at the recent annual meeting of the American Physical Society in New York City. Dr. MacDonald stated that the discovery grew out of analysis of shock waves from the Chilean earthquake of 1960.

An earthquake of such proportions causes everything on the globe to bounce up and down for roughly a month. This bouncing motion amounts to only about a thousandth of an inch, but it registers on sensitive instruments as a simultaneous rising and falling throughout the world. The entire planet expands and contracts in a 20-minute cycle for about 30 days. *AUD*

Spacecraft photographed with 6" telescope by George Adamski, Palomar Gardens, California. Scoutship December 13, 1952.

MYSTERIOUS UFO sightings are not unusual in Antarctica.

LOST CONTINENTS INSIDE THE EARTH

The inside of the earth, as any sixth grader can tell you, is a roiling, liquefied mass of superheated rock and metal; and as such, it bears virtually no similarity to the comfortably solid surface on which we live. But new research by a pair of geologists at MIT has turned up at least one surprising analogy between inner and outer Earth: Parts of the inside, say Thomas Jordan and Kenneth Creager, may feature massive, mobile structures roughly similar to the continents on the surface.

Jordan and Creager borrowed from modern medicine a technique called computerized axial tomography and used CAT scans to peer inside the inner earth in much the same way a neurologist scopes inside a human brain. With this tool they were able to profile the planetwide vibrations set up by large earthquakes; they used that profile to draw three-dimensional maps of the earth's interior.

The maps revealed surprisingly definite areas thousands of kilometers across, in a region that Jordan thinks lies along the boundary between the earth's liquid-iron core and its mantle, the next layer up. Just as our surface continents are chemically distinct from their neighboring layers—denser than the air above but lighter than the rocky lithosphere below—so these inner "continents" seem chemically distinct from both the mantle above and the core below.

Jordan can't yet offer a precise chemical description of the "slag" or "scum" that composes these inner continents. "This is the first time we've tried to do this mapping so deep inside the earth," he says. "I'm sure the earth will have some surprises for us."—Bill Lawren OMNI

'Solid Earth' 'Tain't So, Geophysics Parley Hears

The "solid earth" is far from solid, geophysicists pointed out Wednesday at the regional meeting of the American Geophysical Union here.

The North Pole is sliding toward Greenland; the earth's crust rises and falls with the tide a few inches each day; and Oregon can expect more earthquakes, speakers noted at the opening session.

Scientists generally agree, on the basis of new measurements made with the aid of geodetic satellites that the North Pole is slipping about nine centimeters a year toward Greenland, William Kaula of the Institute of Planetary Sciences, University of California, told delegates to the annual technical conference.

Some scientists think California is sliding seaward, Kaula said, but others think not. Present methods of measurement are not sufficiently exact.

Laser beams may permit more accurate measurements of latitude and longitude when they can be made more powerful, he said. Use of laser beams to track satellites so far has scored "only a few successes," he explained.

Earth's Crust Shifts

Russian scientists have been particularly active in measuring movements of the earth's crust and have cooperated with the U.S. and other nations in measuring tidal effects in the earth's crust, he said.

The earth's crust is moving horizontally northward on the west side of the Hayward Fault in California — 3 feet in the 12 years between 1951 and 1963 — Kaula pointed out.

"This means we can expect an earthquake, a big one, pretty soon. But we don't know when — or why," he concluded.

The Alaska earthquake of March 27, 1964, worst in decades, rotated 40 - mile - long Montague Island 30 seconds of arc — about 30 feet, recent measurements show.

Isle Thrust Up

At the same time a portion of the island was thrust violently upward more than 30 feet.

The theory that whole continents are drifting over the crust of the earth has not been confirmed by actual measurements as yet, Kaula said.

Prof. Joseph Berg of the department of oceanography, Oregon State University, reported on studies now being conducted at Billings, Mont., and elsewhere to devise means of predicting earthquakes in advance.

No predictions are possible as yet, he said, except in a general sense.

Quake Predicted

Stresses building up in the earth's crust indicate Oregon can expect quakes of significant severity, but where and when they will occur is beyond the present state of the art, Prof. Berg told The Oregonian.

Walter C. Sands, University of Washington, reported on Cobb Seamount, the 9,000-foot mountain that lies submerged in the Pacific 270 miles off the coast of Washington, with its head just barely under the surface of the sea. OREGONIAN

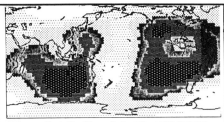

CAT scan of the inside of the earth: Normally used to diagnose diseases, the images above reveal continents inside our planet.

The following five letters illustrate the futility of trying to cooperate with the federal government and institutions of higher learning pertaining to UFOs in general and the Adamski case in particular. Note the lack of a signature on the letter from M.I.T.

DEPARTMENT OF NUCLEAR ENGINEERING

MASSACHUSETTS INSTITUTE OF TECHNOLOGY

77 Massachusetts Avenue   Cambridge, Massachusetts 02139

September 21, 1987

Mr. W.O. Clendenon
P.O. Box 756
Biloxi, MS 39533

Dear Mr. Clendenon:

Thank you for telling us about your work on the ASP project. The concept seems very ingenious but involves a lot of technology that is not being studied in the Nuclear Engineering Department and we will not be able to comment on it. A copy of your material has been forwarded to appropriate people at MIT's Plasma Fusion Center but I am afraid this is somewhat out of their realm of expertise also. I strongly suggest that you send copies of this material to NASA and to the Strategic Defense Initiative Office in Washington; they have already spent a lot of money on power supplies of this type.

Massachusetts Institute of Technology
77 Massachusetts Avenue
Cambridge, Mass. 02139

Mr. W.O. Clendenon

P.O. Box 756

Biloxi, MS 39533

118

DEPARTMENT OF DEFENSE
STRATEGIC DEFENSE INITIATIVE ORGANIZATION
WASHINGTON, DC 20301-7100

EA   January 19, 1988

Honorable Thad Cochran
P.O. Box 22581
Jackson, Mississippi 39225-2581

Dear Senator:

Thank you for your recent letter to Assistant Secretary of
Defense Carlisle concerning Mr. W.D. Clendenon's proposal to the
Strategic Defense Initiative Organization (SDIO). Mrs. Carlisle
forwarded your letter to my office for action.

After a thorough review of the files of unsolicited proposals
contained in SDIO's Contracts and Procurement Office, I have
failed to turn up any record of receiving Mr. Clendenon's
proposal. I realize he has a copy of a registered receipt
indicating acceptance at the Organization, unfortunately, the
signature is unfamiliar to me and cannot be tracked down.

I would appreciate it if Mr. Clendenon would resubmit his
proposal for evaluation by SDIO. He should send the proposal to
the attention of Mrs. Sarah Wilhelm, SDIO Contracts and
Procurement. I apologize for any inconvenience resubmitting the
proposal may cause.

Sincerely,

JOHN C. DEWEY
Captain, USN
Director, External Affairs

DEPARTMENT OF THE NAVY
OFFICE OF THE CHIEF OF NAVAL RESEARCH
ARLINGTON, VIRGINIA 22217-5000

IN REPLY REFER TO
5870
Ser OOCC12
025/028
23 February 1988

Mr. William Dewey Clendenon
Post Office Box 756
Biloxi, Mississippi 39533

Dear Mr. Clendenon:

Your letter of 1 February 1988 and the accompanying documentation entitled "Atomic Aero Steam Power Planetary Gyro Turbine" have been reviewed. Your material does not appear to relate to any present or anticipated areas of Navy research and development, and accordingly, does not warrant evaluation by the Navy.

Your concern in making your ideas available to the Department of the Navy is appreciated, and we regret that we can take no further action on them. We are returning your material for your further use.

Sincerely,

W. F. McCARTHY
Associate Counsel/Senior
OCNR Patent Attorney
By direction of
Chief of Naval Research

Enclosure

W. D. Clendenon
P. O. Box 756
Biloxi, MS   39533

Dear Mr. Clendenon:

It has come to our attention that you have
previously submitted an invention for evaluation. This office has no record of having
received your invention and request that you
resubmit your invention in an unsolicited proposal format as suggested in the attached
guide.

Please forward your proposal to the address
indicated in the guide so that it can be
entered into our unsolicited proposal tracking
system.

Please accept our apologies for any
inconvenience. Your interest in our requirements is appreciated.

Sincerely,

SARA L. WILHELM
Procurement Assistant

DEPARTMENT OF DEFENSE
STRATEGIC DEFENSE INITIATIVE ORGANIZATION
WASHINGTON, DC 20301-7100

EA

9 May 1988

Honorable Thad Cochran
P.O. Box 22581
Jackson, Mississippi 39225-2581

Dear Senator:

Thank you for forwarding Mr. W.D. Clendenon's research paper for consideration by the Strategic Defense Initiative Organization (SDIO). Mr. Clendenon's paper was reviewed by SDIO's Innovative Science and Technology Office for possible application to SDI research projects. Unfortunately, their review of Mr. Clendenon's paper has failed to produce favorable results.

Mr. Clendenon proposed effort is based on the idea that the earth's center is hollow. To our knowledge, there is no scientific evidence that this is true; in fact, the earth's center consists of molten core with volcanos serving as vents. Should Mr. Clendenon have scientific evidence to the contrary, we would be pleased to have him resubmit his idea in proposal form as suggested in the attached Unsolicited Proposal Guide.

If you or Mr. Clendenon should have additional questions concerning the review of his research paper, please do not hesitate to contact me or Mrs. Sarah Wilhelm in our Contracts and Procurement Office.

Sincerely,

JOHN C. DEWEY
Captain, USN
Director, External Affairs

Enclosure: As Stated

OTHER PLANETS

Whether the UFOs are interplanetary spaceships is yet to be proven. The Federal government has yet to prove that UFOs are not real. Determining whether UFOs are real, intelligently-controlled flying machines has been taken into the hands of the public. *Listen up, Congress! For UFOs are still over your heads.*

Adamski publicly stated that the people he encountered flying the UFO Scout Ships informed him that they were from Venus, Mars, and Saturn. However, NASA has informed the public that Venus, for instance, is too hot. NASA has made equally negative remarks pertaining to other planets such as Mars and Saturn. Now there is something wrong here. NASA or Adamski's mysterious friends, perhaps both, appear to be deliberately feeding the public misinformation. What the public wants Congress to do is hold open hearings on UFOs to find out who is telling the truth about UFOs and who is not telling the truth.

Some astronomers and government spokesmen attempt to convince the public that interplanetary UFOs approaching Earth would be detected in advance. But even with I.F.F. (Identification Friend or Foe) detection systems, the argument is full of holes. Government spokesmen indicate our own Stealth Aircraft can penetrate enemy defenses undetected. If our Stealth Aircraft can penetrate enemy defenses undetected, what's to stop interplanetary UFOs from doing the same thing in a more efficient manner? Our astronomers and their telescopes as well as our military I.F.F. anti-aircraft defense systems failed miserably in detecting UFOs on December 7, 1941. The Pearl Harbor UFOs were identified after they attacked, not before, or so we were told.

The Adamski UFO case may turn out to be part of one of the most fantastic cover stories ever devised. The questions are: Who devised the plan and for what reason was a plan needed? The cause for some of the confusion and censorship of the Adamski UFO case must be shared by some high profile individual UFO researchers, private UFO research organizations, various members of the scientific community, some, but not all, of the news media, and many publishers as well as certain agencies and individuals of the United States Government. The bottom line is: the public has been taken for a ride on UFOs.

At this point one is again reminded of SUN TZU: *"One who is confused in purpose cannot respond to his enemy."* In the case of UFOs, is there a public enemy, and if there is, is it the UFOs or governments the world over?

UFO MOTHERSHIPS

Inside the Space Ships told of Adamski's experiences aboard large cigar-shaped UFO motherships. It was said these huge ships were the main means of interplanetary transportation for the mysterious UFOnauts. The large motherships carried small scout craft, stored in their interiors, for use in the atmospheric regions of a planet. To start with, there is not much detailed information to properly research the propulsion systems of the UFO motherships. But even the longest journey starts with one small step.

If form determines function, the first small step toward discovering the Adamski UFO mothership propulsion is the form of the mothership itself, its cigar shape. A round or disc shaped air frame is a more suitable form for the little scout craft. The ability to maneuver in any direction, plus take off and land vertically, is best executed in an aircraft of a circular air frame design. The cigar or missile shaped design of the UFO Mothership would be more efficient for long distance forward flight of spacecraft. The mothership diagrams No. 3 and 10 as illustrated in Adamski's book *Inside the Space Ships* appears to be an Athodyd, technically called a continuous thermal duct. The basic air frame of the Aero-Thermo-Dynamic-Duct is simply a long tube with a divergent forward gas intake and a convergent exit or exhaust nozzle at its stern. We now have an Athodyd in form. But the propulsion for the Athodyd formed UFO mother-ship is wide open for serious UFO researchers to work on. UFO mothership diagrams No. 3 and 10 show the Scout Ship landing chutes (with air locks) topside. The interior Scout Ship hangar deck is placed somewhat amid ship. The Scout Ship launching chute (with air locks) is placed on the bottom side of the motherships. This is a very efficient design for a flying aircraft carrier. The United States Navy employed flying aircraft carriers in the form of dirigibles (cigar-shaped, lighter-than-air aircraft) back in the 1930s. Navy scout planes were launched and retrieved while the mothership dirigibles were air borne. Adamski described similar flight operations aboard the UFO motherships. Interplanetary origins for Adamski's UFO motherships is a number one possibility.

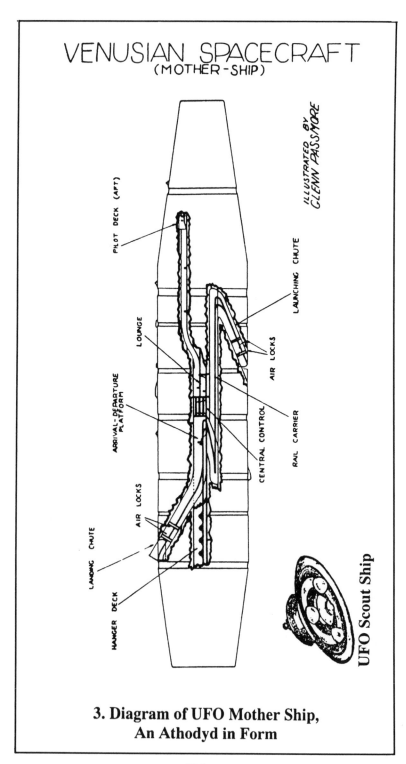

VENUSIAN SPACECRAFT
(MOTHER-SHIP)

ILLUSTRATED BY GLENN PASSMORE

PILOT DECK (AFT)

LAUNCHING CHUTE

LOUNGE

AIR LOCKS

ARRIVAL-DEPARTURE PLATFORM

CENTRAL CONTROL

RAIL CARRIER

AIR LOCKS

LANDING CHUTE

HANGER DECK

UFO Scout Ship

**3. Diagram of UFO Mother Ship,
An Athodyd in Form**

UFO MOTHER SHIP

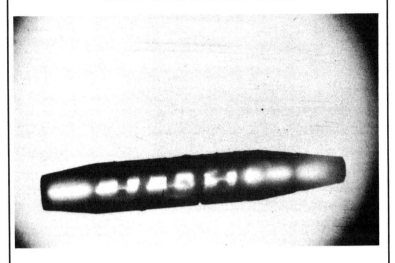

George Adamski, California, 1952

George Adamski, California, 1952

UFO Scout Ships Leaving UFO Mother Ship

This is the last letter Mr. Adamski write ... [handwritten] Madelaine Rodeffer 12905 Falmouth ... age 65 Silver Spring Md. 20904 tel # code 301 622 1007

CRUSADE FOR THE TRUTH ABOUT INHABITANTS OF OTHER WORLDS

EIGHTEEN YEARS OF ALIBIS AND DOUBLE TALK.
BY GEORGE ADAMSKI

Mystery upon mystery has been created over the minds of the people, instead of the truth, of visitors as humans from sister planets.

Even the Space Program has been inaugurated as a result of their coming, with the purpose of us being able to go to their planets. The Air Force has been in charge of this investigation.

The Wright Field, Dayton, Ohio; and the Blue Book, and those whose responsibility it was to investigate reports (like Captain Ruppelt and many others) had never discarded theories, or stated that all the reports were false, but that certain percentages existed which could not be accounted for.

Many false reports were put out during the past eighteen years by the Air Force with the help of astronomers and so-called experts regarding the Unidentified Flying Objects. Most of the alibis were an insult to the intelligence of the American people. If some of the reports of the Air Force were true, why is it then, that they still keep sending Jets up to chase these objects, or continue their research on them?

We are going, or are trying to go, to other planets in our space program. This program was started because of them. This was certain acknowledgement that whoever these people are, and can come from the stars to us, we should be able to go to the stars to them. We cannot deny that this is the very thing we are trying to do. So we must have some concrete information regarding these visitors or we would be shooting for a blind target.

It is about time that this eighteen year mystery should be cleared up by our government, which holds the key to the truth.

A politician is a politician whether he be a Democrat or a Republican, and once elected to office, he wants to remain there. And he will not tell the truth, regardless what kind of truth, if he is not sure of his people, being aware they will not re-elect him to office if he should displease them. Yet he would be willing to tell the truth if he knew that he had the support of his people.

If he did not know the truth regarding the subject, he could demand that truth from such office as is responsible for it, and give it to his people.

So all those that believe in inhabited planets and that we do have visitors from them coming here, should write a letter to his representative and demand the truth.

Should such a representative alibi or send out a pamphlet which the Air Force has prepared to answer such an inquiry, this type of representative is not worth the time or consideration for re-election, even with an alibi that he doesn't know enough about the subject. For he is in a position to have the books opened to him by such a department that is responsible for the collection or gathering of such information, which is the Air Force.

Instead we should concentrate on men and women that we know are not afraid to tell the truth. Take for an example, men like former Governor Knight of California. When he was asked on a TV panel if he believed in flying sacuers he said that he did, that he knew they were there and that we are being visited. He said much more and gave his reasons for it, which were logical ones.

There is a rumor now, that he will be running for Governor of California again, and this is what I mean. It is this type of man that we want to represent us, regardless of the party, men and women that are not afraid to tell the truth.

I believe there are better than 200,000 people in this nation that believe in inhabitants of other worlds, or the flying sacuers. This group, if directed in proper channels, can elect people to office, State or National, men and women that will tell the truth and dissolve the mystery of eighteen years, and prepare our people and the people of the world for what is to come.

Every sincere believer in the flying saucers, regardless of what other opinions he has, should concentrate his efforts on the men of truthful caliber that when elected to office in the future, will tell the truth, men like former Governor Knight. This is the only way and the only hope by which the truth will finally be told to the people.

As everyone knows, no politician wants to be defeated, but instead he wants to maintain the office. And 200,000 votes are not to be sneezed at! And this is a small number. There could be twice that many believers that want the truth. So let us get together. We have never tackled the politicians before and it is about time that we should.

Even President Johnson, prior to becoming Vice President, made a statement which was published in our papers when flying saucers were the news of the day; that he believed in them and he knew they were there. Some of our congressmen have even seen them in flight while journeying in our own planes. The newspaper morgue has these records. Why don't they tell us that today?

People acting in the field of politics have gotten good results for the good of the nation once they organized and put the pressure upon the representatives. Let us do it now for the good of our civilization.

SUPPLEMENTARY
DOCUMENTATION

**Censorship: UFOs
Under Federal Orders
vs.
UFO: Under Freedom Of
Information Act**

"The issue is the right to know."
W.D. Clendenon

Office Memorandum · UNITED STATES GOVERNMENT

TO : Director, FBI (100-364606) DATE: December 15, 1953

FROM : SAC, San Diego (100-8382)

SUBJECT: GEORGE A. ADAMSKI, aka Professor ADAMSKI
George A. Adamski, George A. Adamsky
SECURITY MATTER - C

Re Los Angeles letter to Bureau, 12/10/53, San
Diego letter to Bureau, 9/22/52, and telephone calls from
Bureau to San Diego, 12/14 and 15/53.

Re San Diego letter to Bureau, 9/22/52, advises
the Bureau that no additional investigation was being conducted relative to the subject.

For the information of the Bureau and the Los
Angeles Office, the subject owns and operates the Palomar
Gardens Cafe located five miles east of Rincon, California
at a point where the highway branches off leading to the
Mt. Palomar Observatory.

He is an amateur astronomer and for the past
several years this office has received complaints relative
to the subject's having seen flying saucers in the vicinity
of his establishment. He exhibits photographs purported to
be of flying saucers to patrons of his establishment. OSI
of the Air Force has done considerable investigation relative to these complaints and lends no credence to the truth-
fulness of ADAMSKI's statements.

The Bureau's attention is directed to San Diego
letters dated 1/28/53 and 3/23/53 relative to this matter.
For the information of the Los Angeles Office, the following
signed statement was taken from ADAMSKI on 3/17/53:

COPIES DESTROYED   "Palomar Gardens
41 JUL 10 1963   March 17, 1953

"TO WHOM IT MAY CONCERN:
                                                    (b)(7)(c)
    "I, George Adamski, hereby make the following
signed statement to ▮▮▮▮▮▮▮▮▮ Agent F. B. I. and
▮▮▮▮▮▮▮ and ▮▮▮▮▮▮ Agents of the Air Force. This
statement is voluntary and no threats or promises have been
made to me.

    ▮▮▮: CS   RECORDED SF-111 100 - 3 75 72

AIRMAIL REGISTERED
2 - Los Angeles (100-24442)(Registered)   DEC 17 1953

130

SD 100-8382

        "I understand that the Federal Bureau of Investi-
gation and the United States Air Force investigate complaints
affecting the security of the United States. I understand
that they make no recommendations as to the validity or
non-validity of these complaints.

        "I have not and do not intend to make statements
to the effect that the U. S. Air Force or Federal Bureau of
Investigation have approved material used in my speeches.

        "I have read the above statement and it is true
and correct to the best of my knowledge.

                        "Signed,

                        "/s/ GEO_ ADAMSKI

"GA:lm   "George Adamski
                        Star Route
                        Valley Center, Calif.

"Witness:

"/s/ 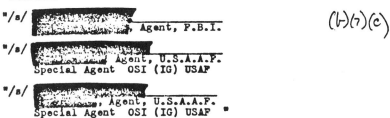, Agent, F.B.I.   (b)(7)(c)

"/s/ _____, Agent, U.S.A.A.F.
      Special Agent OSI (IG) USAF

"/s/ _____, Agent, U.S.A.A.F.
      Special Agent OSI (IG) USAF "

        As instructed by the Bureau, ADAMSKI will be con-
tacted in the immediate future, at which time he will be
requested to cease and desist making any reference to the
FBI in his talks or in any publications which he might issue.
The Bureau will be advised of the results of this contact.

Office Memorandum • UNITED STATES GOVERNMENT

TO : Mr. Tolson   DATE: Dec. 16, 1953

FROM : L. B. Nichols

ALL INFORMATION CONTAINED HEREIN IS UNCLASSIFIED DATE 5/11/82 BY SP/mn/rw

SUBJECT: GEORGE A. ADAMSKI
SECURITY MATTER - C
COAUTHOR OF "FLYING SAUCERS HAVE LANDED"

(b)(7)(c)

SYNOPSIS:

George A. Adamski, 62, described as mentally unbalanced and a "crackpot" of Palomar Gardens, Valley Center, California, is coauthor with one Desmond Leslie of the book "Flying Saucers Have Landed," first printed in September, 1953, and now in its fifth printing. Adamski claims to have taken trip around Moon in an interplanetary ship piloted by men from Venus. A 3-13-53 article in the Riverside, California, "Enterprise" states Adamski, in a speech on space travel befor the Corona, California, Lions Club 3-12-53, stated his material had all been cleared with FBI and Air Force Intelligence. On 3-17-53, Adamski interviewed by SA ▬▬▬▬ San Diego Office, ▬▬▬▬ and ▬▬▬▬ Office of Special Investigations agents. Adamski denied making statement, wrote letter to the editor correcting record, and gave signed statement witnessed by the three Agents to effect he had not and did not intend to make statements that the FBI and OSI approved his material. He said he understood these agencies make no recommendations. Copy of statement demanded by and given to Adamski 3-17-53. By letter dated 12-11-53, to Director from ▬▬▬▬ Better Business Bureau, Los Angeles, California, ▬▬▬▬ states Adamski exhibited a document signed by three Agents purporting to clear his material. ▬▬▬▬ on 12-10-53, advised Los Angeles Office the Better Business Bureau intends to label Adamski's book as a fraud. Air Force officers in Pentagon likewise received letter from ▬▬▬▬ dated 12-11-5. SAC Willis at San Diego instructed to have Agent, accompanied by OSI, read riot act to Adamski in no uncertain terms, diplomatically retrieve copy of signed statement if possible admonish him for statements and false representations. Proposed letter to Better Business Bureau prepared. WILLIS TO ADVISE AS TO RESULTS.

by telephone on 12-14-5

(b)(7)(c)

RECOMMENDATION:

That the attached letter to Mr. ▬▬▬▬ Research Division, Better Business Bureau, Los Angeles, California, be forwarded pointing out true facts of the FBI's relationship with Adamski and fact this Bureau has not endorsed, approved, or cleared Adamski's speeches or book.

Attachment
cc - Mr. Ladd
cc - Mr. Belmont
cc - Mr. Jones

RECORDED-86

CLARENCE CARPENTER, 608 South Main, Corona, California, reporter of the story on ADAMSKI advised the statement, "his material had all been cleared with the Federal Bureau of Investigation and Air Force Intelligence," was a direct quote from his notes.

Unless advised contrary by Bureau no further investigation will be conducted.

Office Memorandum • UNITED ! A'TS GOVERNMENT

TO : Director, FBI(100-26606) AIR MAIL   DATE: 12-24-53

FROM : SAC, San Diego (100-8382)

SUBJECT: GEORGE A. ADAMSKI, WAS
SECURITY MATTER - C

Remytel 12-17-53 in the above-captioned matter, calling attention to a contact made with the captioned individual by two Agents of this office 12-17-53, at which time Mr. ADAMSKI was admonished for having stated in public utterances that his remarks with reference to flying saucers were with the approval of the FBI.

In this regard, I am quoting below a letter dated 12-22-53 addressed to Special Agent ███████████ by ███████████ an attorney, Beverly Hills, California, which is self-explanatory: (b)(7)(c)

"Dear Mr. ████

"   I have been consulted by GEORGE ADAMSKI of Palomar Gardens, California. On December 17th Mr. ████, Mr. ████ and yourself called on Mr. ADAMSKI relative to a radio broadcast, certain documents and letters concerning his activities in observing unidentified flying objects. (b)(7)(c)

"   I am writing you to ascertain if there are any pending legal proceedings wherein it will be necessary that Mr. ADAMSKI have legal counsel. I have advised Mr. ADAMSKI that possibly there could be a question of security involved and a responsibility on his part if references were made to government agencies in an attempt to give credibility to his material or himself as an author.

"   I will be present with Mr. ADAMSKI at any hearing if it is necessary and wish to assure you and your Bureau that if I appear as Mr. ADAMSKI'S counsel it will be on the basis of cooperation.

"   May I hear from you at your earliest convenience?

COPIES DESTROYED
41 JUL 10 1963   (b)(7)(c)

SA ████ was one of the Agents who interviewed Mr. ADAMSKI, and he informs me that in their talk with the subject they admonished him in strong language and gave him reason to believe that, if he did not cease his references to the FBI, consideration would be given to referring his

RECORDED - 26   100-395-23-10

INDEXED - 26   DEC 28 1953

ALL INFORMATION CONTAINED
HEREIN IS UNCLASSIFIED
DATE ██████ BY ██████

JAN 13 1954

SD 100-8382

activities to the USA'S Office. It appears that Mr. ADAMSKI has secured the (b)(7)(C)
advice and counsel of Mr. ███ for fear some legal action might be con-
templated.

UACB by 12-30-53, the letter of Mr. ███ will be acknowledged in the following
language: (b)(7)(C)

"Dear Mr. ███

"   Special Agent ███ has called to my attention
your letter addressed to him dated December 22, 1953.   (b)(7)(C)

"   It is noted you have been retained by Mr. George Adamski of
Palomar Gardens, California, and as his attorney you are desirous of knowing
whether there are any pending legal proceedings with reference to your client.
As you may be aware, the functions of the Federal Bureau of Investigation are
purely investigative in nature, and all matters relating to the prosecution of
cases within our jurisdiction are handled by the United States Attorney's Office.

"   I am placing your letter on file and appreciate your offer
to be of assistance.

                                        Very truly yours,

                                        No further
                                        action necessary.
                                        WW, 1-11-54

135

EXPERIMENTAL AIRCRAFT ASSOCIATION

A Non Profit Organization Dedicated to the Advancement of Home Built Aircraft and Private Aviation

Paul H. Poberezny, President   Harry Zeisloft, Vice-President   Robert E. Nolinske, Sec.-Treas.

Phone Garden 5-4860

9711 W. Forest Park Drive
Hales Corners, Wisconsin

17 August 1959

William D. Clendennon, Jr.
407 Highland Park St.
Lebanon, Tenn.

Dear Sir:

Received your letter of July 5 and am sorry with the
delay in answering but my mail here at Headquarters
mounts up.

It was quite an interesting letter you sent relative
to the foreign objects. Regarding circular wing
aircraft I would suggest that you write to Al Loedding
4324 Schrubb Dr., Dayton 29, Ohio. Al Loedding is
very interested in circular wing aircraft and is also
an aerodynamist for the Air Force at Wright Field
and is in charge of the wind tunnel. He is also vice
president of the Dayton, Ohio area chapter.

I can certainly understand your problem relative to
your patents and as I always say, "There's no money
in being a pioneer". It always seems that after someone does the pioneer work, someone else cashes in on
it.

Sincerely

Paul H. Poberezny
President

BOARD OF DIRECTORS

PAUL POBEREZNY   ROBERT NOLINSKE   HAROLD GALLATIN   GEORGE HARDIE, Jr.
HARRY ZEISLOFT   STANLEY J. DZIK   GEORGE GRUENBERGER   KEITH F. KUMMER

EDITORIAL STAFF

PAUL H. POBEREZNY   GEORGE HARDIE, Jr.   ROBERT NOLINSKE   S. H. SCHMID
Editor in Chief   Managing Editor   Associate Editor   Advertising Manager

ATTEND THE 7th ANNUAL FLY-IN AUG. 6, 7, 8, 9, 1959—GREATER ROCKFORD AIRPORT. ROCKFORD. ILLINOIS

February 26, 1962

Dear Mr. Clendenon:

Many thanks for permitting me to examine the enclosed material.

Having little technological training, I was not able to make a personal evaluation of the matter. Therefore, I showed it to university men who could be considered experts in their realms of electronics and physics. They were of the opinion that your work was unique and possibly of an unprecedented nature. They were also of the opinion that you were dealing with theories with which they were unfamiliar — and therefore unqualified to judge.

Somewhere, sometime, someone will discover a workable process of nullifying gravity and of putting it to work. Whether you have done it, I cannot say, but I hope you do, in case you haven't already.

Sincerely yours,
Frank Edwards

Over the years, I exchanged research information on UFOs with Frank Edwards. Like many other famed UFO researchers in Frank's time period, he did not think highly of Adamski. In fact Frank referred to Adamski's UFO Scout Ship photo of Dec., 1952 as "the top off a 1937 vacuum cleaner." I sent Frank some research papers on my Adamski Scout Ship photo analysis so that he might have them evaluated professionally. He replied in his letter of Feb. 26, 1962.

Note: Autographed copy to me from Frank Edwards.

FLYING SAUCERS·SERIOUS BUSINESS

BY FRANK EDWARDS

OVERWHELMING NEW EVIDENCE THAT THEY ARE REAL!!!!!!!!!!!

THE BOOK
THAT SMASHES
THROUGH
THE BARRIER OF
OFFICIAL SILENCE
WITH THE
EXCLUSIVE
STORY!

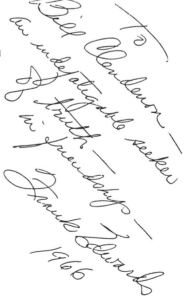

OFFICIAL BUSINESS

DEPARTMENT OF STATE
WASHINGTON

Prof. George Adamski
Star Route,
Valley Center
California

My Dear Professor:

For the time being, let us consider this a personal letter and not to be construed as an official communication of the Department. I speak on behalf of only a part of our people here in regard to the controversial matter of the UFO, but I might add that my group has been outspoken in its criticism of official policy.

We have also criticized the self-assumed role of our Air Force in usurping the role of chief investigating agency on the UFO. Your own experiences will lead you to know already that the Department has done its own research and has been able to arrive at a number of sound conclusions. It will no doubt please you to know that the Department has on file a great deal of confirmatory evidence bearing out your own claims, which, as both of us must realize, are controversial, and have been disputed generally.

While certainly the Department cannot publicly confirm your experiences, it can, I believe, with propriety, encourage your work and your communication of what you sincerely believe should be told to our American public.

In the event you are in Washington, I do hope that you will stop by for an informal talk. I expect to be away from Washington during the most of February, but should return by the last week in that month.

Sincerely,

R. E. Smith
Cultural Exchange Committee

138

A.

B.

C.

D.

Photographs A, B, and C above are reproductions of a series of postage stamps issued by the Caribbean Island of Grenada. Each stamp shows an illustration of a sighting or a reproduction of a photograph taken on or about the date shown on the stamps. Picture "A" shows the familiar "bell type" scout craft as photographed by George Adamski on December 13th 1952, while stamp "B" shows an illustration of "lights" in front of the moon much like a photograph also taken by George Adamski with a 6" telescope during the early 1950's. Grenada's prime minister Sir Eric Gairy was instrumental in the push for what he calls an urgent need for "research into Unidentified Flying Objects." He carried the message to the United Nations General assembly.

Photograph "D" is a reproduction of a postage stamp issued by the Republic of Equatorial Guinea. It also shows a "ball type" scout craft as Adamski had photographed many times.

E.

Above (photo "E") is the photograph refered to above as taken by George Adamski in the early 1950's. All photographs on this page are available from the enclosed address.

CLARK PUBLISHING COMPANY

845 CHICAGO AVENUE, EVANSTON, ILLINOIS 60204
TELEPHONE: 869-2550 AREA CODE 312

March 19, 1964

Mr. William D. Clendenon, Jr.
P. O. Box 926
Portland, Oregon

Dear Mr. Clendenon:

I do appreciate the long letter you wrote me on February 23 and
the information on your theories of UFO propulsion. Unfortunately,
we do not have anyone on our staff who is enough of an electronics
expert to comment intelligently on your theories, but I will scout
around and see who I might have check them out. Meanwhile, it
was nice of you to think of us.

Now as to some of the questions you have asked, I will try to
answer them to the best of my ability. As far as I know, we do
not have photographs or information of John W. Keely or the book
by Mrs. Bloomfield Moore that you refer to, nor do we have
Leonard Cramp's book "Space, Gravity and the Flying Saucers".
However, for Leonard Cramp's book I suggest that you write to
Mr. Max Miller, 1420 S. Ridgeley Drive, Los Angeles, California,
or I suggest that you correspond with James W. Moseley, editor
and publisher of "Flying Saucer News", who can be reached at
119 East 96th Street, New York 28, N.Y. P O Box 163 Ft Lee NJ.

You wonder why there has not been an open congressional hearing
on the subject of UFO's, and I can only tell you that Donald Kehoe
has been trying to get Congress to investigate them for several
years, without any luck.

I hope that something comes of my sending your material to other
persons. It may be of some help to you.

Very truly yours,

Curtis G. Fuller
Publisher

CF:sw

P.O.Box 926
Portland, Oregon 97207 , U.S.A.
June 1, 1965

De Havilland Canada
Montreal, Canada

To whom It May Concern:

I am writing your Company in regard to my invention of an aircraft
propelled by an electro-magnetic motor, utilizing a mercury ballast
for directional control, etc. I would like to know in what way I
could disclose my invention to your Company or your Government as
well as other interested concerns in Canada, and still protect my
inventor's rights in case disclosures and further developments were
to take place within Canada, such as patent rights, etc. Or could
you possibly put me in contact with someone representing your
Government or Company who would be qualified to observe my findings
and discuss it from that point on.

An early reply would be greatly appreciated.

Sincerely yours,

William D. Clendenon, Jr.

Paul: This came to us by dog sled, no doubt
because of the mailing address. Would you
like to acknowledge.

A.J.W.

141

CLARK PUBLISHING COMPANY

845 CHICAGO AVENUE, EVANSTON, ILLINOIS 60204
TELEPHONE: 869-2550 AREA CODE 312

June 9, 1965

Mr. William D. Clendenon, Jr.
P. O. Box 926
Portland, Oregon 97207

Dear Mr. Clendenon:

I regret more than I can say that your letter of February 15th got mislaid in our file and only turned up today, so I am hastening to answer it.

As you know, a great deal of your inquiry is academic because of the death of George Adamski, but as for Adamski's criticism of FATE magazine I would like to make the following comments:

1. We believe Adamski was a fraud.

2. He was angry with us because we refused to publish material that we believed to be fraudulent.

3. We were acquainted with Adamski long before he ever published a book and, in fact, published an early article of his and because of events connected with it became convinced that he was either a conscious or an unconscious fraud and would have nothing more to do with him.

4. It is for this reason and this reason alone, so far as I know, that he chose to talk about us from time to time.

5. As far as I know we have not published anything on your research -- in fact, if we had published anything on it you would have been informed and we would have paid you.

I hope this answers your questions.

Very truly yours,

Curtis G. Fuller
President

CGF:e

J. W FULBRIGHT, ARK., CHAIRMAN

JOHN SPARKMAN, ALA.   BOURKE B. HICKENLOOPER, IOWA
MIKE MANSFIELD, MONT.   GEORGE D. AIKEN, VT.
WAYNE MORSE, OREG.   FRANK CARLSON, KANS.
RUSSELL B. LONG, LA.   JOHN J. WILLIAMS, DEL.
ALBERT GORE, TENN   KARL E. MUNDT, S DAK
FRANK J LAUSCHE, OHIO   CLIFFORD P. CASE, N.J.
FRANK CHURCH, IDAHO
STUART SYMINGTON, MO.
THOMAS J. DODD, CONN
JOSEPH S. CLARK, PA
CLAIBORNE PELL, R.I.
EUGENE J. MC CARTHY, MINN.

        CARL MARCY, CHIEF OF STAFF
        DARRELL ST CLAIRE, CLERK

United States Senate

COMMITTEE ON FOREIGN RELATIONS

July 6, 1965

Mr. William D. Clendenon, Jr.
P. O. Box 920
Portland, Oregon 97207

Dear Mr. Clendenon:

This will acknowledge your letter of June 25 in regard to the reply you received from the Superintendent of the Arlington National Cemetery concerning the interment of Mr. George Adamski.

In view of the confusion created by the Superintendent's reply, I requested, and have received, a report from his office on this matter. The report is enclosed for your consideration. As you will note, the inaccurate response of the Superintendent to your letter was apparently due to a clerical error. The information contained in my letter of June 7 is correct.

I am glad that I have been able to assist you. If you should need help in other matters in the future, I shall be pleased to hear from you again.

With best regards,

Sincerely,

Wayne Morse

WM:sdm
Enclosure

143

DEPARTMENT OF THE AIR FORCE
WASHINGTON

OFFICE OF THE SECRETARY

9 JUL 1965

Dear Mr. Clendenon:

The President has asked that I acknowledge your recent letter to him.

The matter about which you wrote has been placed in the hands of appropriate Air Force officials for their careful consideration. You will receive a substantive reply to your letter within the next several days.

Sincerely,

THOMAS G. CORBIN
Major General, USAF
Director
Legislative Liaison

Mr. William D. Clendenon, Jr.
Box 926
Portland, Oregon 97207

144

SUD-AVIATION

SOCIÉTÉ NATIONALE DE CONSTRUCTIONS AÉRONAUTIQUES
SOCIÉTÉ ANONYME AU CAPITAL DE 248.800.000 F.

SIÈGE SOCIAL
37, 8e av MONTMORENCY
— PARIS·16e —
R. C. SEINE 57 B 8481
NO D'ENTR. 270·70·116·0.088

DIRECTION GÉNÉRALE

TÉLÉPH : 224-84-00
(30 LIGNES GROUPÉES)
TÉLÉGRAMMES
AEROBUDEO·PARIS
TÉLEX Nᵒ D'APPEL : 20646

V/Réf :
N/Réf : LT/KV/SG - DT/PI - n° 7063

Le 12 Juillet 1965

Mr William D. CLENDINON Jr
P.O Box 926
PORTLAND, OREGON 97207
(U.S.A)

Objet : Votre proposition d'idée

Monsieur,

Nous avons bien reçu votre lettre du 25 Juin 1965 nous proposant de nous
intéresser à un projet de votre invention concernant un avion propulsé
par un moteur électro-magnétique et pouvant décoller et atterrir vertica-
lement.

Nous vous informons tout d'abord qu'en vue de bien respecter les droits
de propriété des inventeurs sur les nouveautés industrielles, il est d'usage
dans notre Société de n'examiner leurs propositions que lorsqu'elles sont
couvertes par des brevets d'invention français et d'après les textes et des-
sins de ces brevets, ce qui a de plus l'avantage de bien définir l'apport
technique éventuel des parties. Si votre idée est protégée par un brevet
français, vous pourrez nous l'indiquer et nous ne manquerons pas de l'examiner, sans engagement de notre part.

Veuillez agréer, Monsieur, l'expression de nos sentiments distingués.

SUD-AVIATION
Direction Générale

L TRANNOY
Propriété Industrielle

July 26, 1965

R Y A N

Mr. William D. Clendenon, Jr.
P. O. Box 926
Portland, Oregon 97207

Dear Mr. Clendenon:

Thank you for bringing your invention to our attention.
Your invention may have patentable merit; however, it has been
our experience that it is a perilous journey from conception to
the issuance of letters patent. Therefore, Ryan cannot further
consider your invention until a patent application has been filed
with the U. S. Patent Office. If at that time you would care to
submit a copy of the patent application, Ryan would be willing to
evaluate your concept. It is Ryan policy, however, that the sub-
mission of any unsolicited data be accompanied by the enclosed
waiver.

Very truly yours,

RYAN AERONAUTICAL CO.

Wm. M. Flenniken
Legal Assistant

WMF:gf

Encl.

THE DE HAVILLAND AIRCRAFT OF CANADA, LIMITED
DOWNSVIEW, ONTARIO

CABLES
"MOTH" TORONTO

SHIPMENTS
WEST TORONTO

September 7, 1965.

Mr. W. D. Clendenon, Jr.,
P.O. Box 926,
Portland, Oregon 97207,
U. S. A.

        Re: Invention of an Aircraft Propelled
             by an Electro-Magnetic Motor

Dear Mr. Clendenon:

        Your letter dated June 1st., 1965 regarding the above
invention has come to my attention. I must apologize for the delay
in replying to you but for a variety of reasons your letter only
recently arrived on my desk.

        You are to be congratulated for your interest and
inventiveness in the field of aeronautics. However, I must inform
you that due to prior commitments on both Civil and Military projects
the Research Engineering and Development Department of our firm are
fully occupied. Further we cannot undertake discussion of your
invention now or in the near future. Consequently I am returning
herewith the letter you sent us dated June 1st., 1965 and request
that you make no further disclosure of your invention to us.

        Might I suggest you contact: Canadian Patents and
Development Limited, National Research Council, Building M 58,
Montreal Road, Ottawa 7, Ontario, Canada. This Company was formed
by the Canadian Government to develop and licence inventions.
They may be able to advise you or refer you to another aircraft
company or Government Organization better able to be of assistance.

        Yours very truly,

P. Church

        P. Church,
        Assistant to Vice-President
        Research and Planning.

Encl.

THE **BOEING** COMPANY

HEADQUARTERS OFFICES · P.O. BOX 3707 · SEATTLE, WASHINGTON 08124

March 16, 1966
IN REPLY REFER TO
1-2310-7-460

Mr. William D. Clendenon Jr.,
P. O. Box 926
Portland, Oregon 97207

Dear Mr. Clendenon:

We are returning and have enclosed herewith
the Adamski photograph you requested in your
letter of March 6, 1966.

Thank you once again for your interest in The
Boeing Company.

Very truly yours
THE BOEING COMPANY

Leo Siegenthaler
Patent Staff

Enclosure

148

J. W. FULBRIGHT, ARK., CHAIRMAN

JOHN SPARKMAN, ALA.   BOURKE B. HICKENLOOPER, IOWA
MIKE MANSFIELD, MONT.   GEORGE D. AIKEN, VT.
WAYNE MORSE, OREG.   FRANK CARLSON, KANS.
RUSSELL B. LONG, LA.   JOHN J. WILLIAMS, DEL.
ALBERT GORE, TENN.   KARL E. MUNDT, S. DAK.
FRANK J. LAUSCHE, OHIO   CLIFFORD P. CASE, N.J.
FRANK CHURCH, IDAHO
STUART SYMINGTON, MO.
THOMAS J. DODD, CONN.
JOSEPH S. CLARK, PA.
CLAIBORNE PELL, R.I.
EUGENE J. MCCARTHY, MINN.

CARL MARCY, CHIEF OF STAFF
DARRELL ST. CLAIRE, CLERK

United States Senate

COMMITTEE ON FOREIGN RELATIONS

March 16, 1966

Mr. William D. Clendenon, Jr.
P. C. Box 926
Portland, Oregon 97207

Dear Mr. Clendenon:

Thank you very much for your letter of February 22. I appreciate the confidence in me which was implied by your communicating with me again on the subject of your invention VTOL aircraft.

In my opinion, you are following the proper course of action in taking up this matter with the Boeing Company. If I thought that it would serve any useful purpose to discuss this subject further with the Department of the Air Force, or any other branch of the military, I would gladly cooperate. However, I have reviewed again the letter of October 18, supplied by the Department of the Air Force, as well as its October 7, 1965, letter and I feel that everything that could possibly be done on your behalf at the Federal level has been undertaken. As you will recall, the October 7 letter of the Air Force concluded with this:

"Our technical personnel would be pleased to give Mr. Clendenon's invention a thorough analysis; however, we cannot do this until he submits the necessary material in the manner which we outlined to him."

I wish I could relay news of a more encouraging nature, but I know you wish a frank appraisal of the situation as I understand it.

Several of the items you called to my attention earlier are returned for your file because I assume you wish to have them.

With kindest regards,

Sincerely,

Wayne Morse

WM:slb
Enclosures

149

Aerial Phenomena Research Organization

3910 EAST KLEINDALE ROAD
602—326-0059
TUCSON, ARIZONA — 85716

30 April 1966

William D. Clendenon Jr.
P. O. Box 926
Portland, Oregon   97207

Dear Mr. Clendenon:

Apparently we did not receive your first letter.

I have also heard the Adamski ship called a surgical lamp but am not able to help you on that point.

We are enclosing a membership application blank and brochure and a UFO report form.

Please forgive the brief letter- we are swamped with reports, correspondence and investigations, as usual.

Sincerely Yours,

L. J. Lorenzen
DIRECTOR.

enclosures: UFO report form
            APRO Brochure
            Application blank

NOTE:
AIR FORCE
POST MARK?

I DO NOT KNOW
HOW AIR FORCE
POST MARK CAME
TO BE ON MY
DOCTOR BILL.

PHYSICIANS AND SURGEONS
1440 S. W. TAYLOR STREET
PORTLAND, OREGON 97205

Bx 49 Port angeles Wash. 98362

Mr. William Clendenon
4415 SW Primrose
Portland, Ore.

UNIVERSITY OF CALIFORNIA, BERKELEY

BERKELEY · DAVIS · IRVINE · LOS ANGELES · RIVERSIDE · SAN DIEGO · SAN FRANCISCO SANTA BARBARA · SANTA CRUZ

COLLEGE OF ENGINEERING
DEPARTMENT OF CIVIL ENGINEERING
DIVISION OF HYDRAULIC AND
  SANITARY ENGINEERING

BERKELEY, CALIFORNIA 94730

July 26, 1966

Mr. Clendenon
P. O. Box 49
Port Angeles, Washington

Dear Mr. Clendenon:

In reply to your letter of July 14 let me say that I am
not inclined to doubt your statements and observations, but
do feel that your speculations are premature. The information that anyone has regarding the future of propulsion and
energy systems is so limited that speculation on the details
of how advanced systems might be made to operate seems to me
to be out of place. It may be useful to speculate in general
terms on where technology could be taking us, but a reduction-
to-practice, which is what an invention really is, must await
more knowledge.

Sincerely,

J. A. Harder
Associate Professor of
Hydraulic Engineering

JAH:mi

cc: Mr. James Lorenzen

P.S. I am returning your drawings; be assured that I
will keep your information entirely confidential.

152

COSMIC SCIENCE

Study and Application of Universal Laws

Non-Sectarian   Non-Political

~~STAR ROUTE VALLEY CENTER CALIFORNIA~~

~~GEORGE ADAMSKI~~-Deceased
Advisor

New Address
P. O. BOX 2431
FULLERTON, CA 92633

December 29, 1966

Bill Clendenon
P. O. Box 49
Port Angeles, Wash. 98362

Dear Mr. Clendenon:

I no longer carry any of Mr. Adamski's pictures. Shortly before he died I returned his negatives and I have no way of having any pictures made now. I can only suggest you write to Mrs. Alice K. Wells, 314 Lado de Loma Dr., Vista, Calif., and ask her if she has any for sale. She was Mr. Adamski's housekeeper and may have the negatives.

As far as the other photos go, I never have sold these nor do I have a source for them. Color Control Company in Hollywood, Calif. 90028(o311 Yucca) advertizes slides of many ufo pictures but I don't know if they have prints for sale. You might try them and see. They want 50¢ for a catalog or they won't answer their mail.

Sorry I can't help you further.

Sincerely,

C. A. Honey
C.A. Honey

C. A. Honey
P. O. BOX 2431
FULLERTON, CA. 92633

Omak, Washington
23 August, 1967

Mr. Bill Clendenon
P.O. Box 49
Port Angeles, Washington.

Dear Mr. Clendenon:

Thank you for your letter of the 21st and for letting
us see the News item which will be herewith enclosed
as per your request.

Because of reasons that we do not care to discuss, but
which I believe that you will understand, we do not
wish to have the experience my wife had with people
supposed to be from outer space publicised.

It is true that she had a very wierd experience unless
one can believe that she actually was contacted by
persons from another planet, but the experience was
some years ago and nothing of a similar nature has
occured since. We feel that it is best we try to
forget the whole affair.

If you are ever in this area we would like to have
you call to see us. Thanks again.

Yours truly,

Bert Wassell

February 28,1966

Dear William Clendenon:
                    Yours of the 23rd received. Thank you
for your confidence. I only wish that you could have that
much confidence in your own impressions that enabled you to
draw the design for a craft to travel interplanetary space.

                    How can you have any doubts about George
Adamski's experiences after John Glenn Jr., made his memorable
orbit around the Earth and reported seeing " fireflies in space"
which George Adamski described many years before in his book
Inside The Space Ships on page 76. Had G.A. not have traveled in a
space ship - how could he have known what was in outerspace ?

                    Enclosed is a March 1966 Cosmic Bulletin
that might interest you.

                    May I wish you every success.

                                        Sincerely

                                        Alice K. Wells

154

DEPARTMENT OF THE AIR FORCE
HEADQUARTERS FOREIGN TECHNOLOGY DIVISION (AFSC)
WRIGHT-PATTERSON AIR FORCE BASE OHIO 45433

REPLY TO
ATTN OF: TDET/UFO

NOV 1 1967

SUBJECT: UFO Observation

TO:
Bill Clendenon
P.O. Box 49
Port Angeles, Washington 98362

Reference your unidentified observation. The information which we have received is not sufficient for a scientific investigation. Request you complete the attached FTD Form 164 and return it in the envelope provided. Thank you for reporting your observation to the Air Force.

STANLEY C. MANATT, Colonel, USAF
Director of Technology and Subsystems

1 Atch
FTD Form 164 w/envelope

155

LLOYD MEEDS
2D DISTRICT, WASHINGTON

AL SWIFT
ADMINISTRATIVE ASSISTANT

JIM MICHEL
DISTRICT OFFICE DIRECTOR
ROOM 301
FEDERAL BUILDING
EVERETT, WASHINGTON
ALPINE 9-2233

COMMITTEE ON EDUCATION
AND LABOR

SUBCOMMITTEES:
GENERAL EDUCATION
SELECT EDUCATION
SELECT LABOR

COMMITTEE ON INTERIOR
AND INSULAR AFFAIRS

SUBCOMMITTEES:
INDIAN AFFAIRS
IRRIGATION AND RECLAMATION
PUBLIC LANDS

Congress of the United States
House of Representatives
Washington, D.C. 20515

January 29, 1968

Mr. Bill Clendenon
P. O. Box 49
Port Angeles, Washington 98362

Dear Mr. Clendenon:

The Air Force has replied to my inquiry based on your letters. Since NASA does not study UFO's, no inquiry with them would be practical.

Unfortunately, the Air Force cannot provide copies of the photographs you request, as they have never received them. Their analyses of these pictures are based on reprints.

I note that either McLean's or your source reversed the negative in reprinting the photograph you sent. As you requested, it is enclosed.

In any event, I hope the enclosed will help to answer your questions.

Sincerely,

Lloyd Meeds
Member of Congress

LM:eme

enclosure

LLOYD MEEDS
2D DISTRICT, WASHINGTON

AL SWIFT
ADMINISTRATIVE ASSISTANT

JIM PRICE
DISTRICT OFFICE DIRECTOR
ROOM 201
FEDERAL BUILDING
EVERETT, WASHINGTON
ALPINE 9-2233

Congress of the United States
House of Representatives
Washington, D.C. 20515

COMMITTEE ON EDUCATION
AND LABOR

SUBCOMMITTEES:
GENERAL EDUCATION
SELECT EDUCATION
SELECT LABOR

COMMITTEE ON INTERIOR
AND INSULAR AFFAIRS

SUBCOMMITTEES:
INDIAN AFFAIRS
IRRIGATION AND RECLAMATION
PUBLIC LANDS

May 1, 1969

Mr. William D. Clendenon, Jr.
P. O. Box 49
Port Angeles, Washington 98362

Dear Mr. Clendenon:

   Thank you for writing again and enclosing photostats of the return receipts from the material you sent to
NASA in November and January.

   I am sorry that we passed on some misinformation
on our previous inquiry. They have now found your sub-
mittal and have promised me a written evaluation shortly.
I have been unable to discover just why this process has
taken the time it has, although I know the volume of inventions and ideas submitted to them is tremendous.

   At any rate, I hope to be sending you their report
in the near future.

                        Sincerely,

                        Lloyd Meeds
                        Member of Congress

LM:tww

157

DEPARTMENT OF TRANSPORTATION
FEDERAL AVIATION ADMINISTRATION

WASHINGTON, D.C. 20590

JAN 29 1970

Honorable Lloyd Meeds
House of Representatives
Washington, D. C. 20515

Dear Mr. Meeds:

This is in reply to your letter of 16 January 1970 requesting assistance
for Mr. Bill Clendenon of your district in obtaining a possible research
grant and technical information on his mercury-vapor turbine development
for aircraft.

The idea Mr. Clendenon suggests is in the area of basic research which
is not within the responsibilities of the Federal Aviation Administration.
We made inquiries of other government agencies and suggest that the most
proper agency to handle this request would be the NASA Lewis Research
Center.

The NASA headquarters propulsion group identified Mr. R. J. Webber,
Mission Analysis Branch, Advanced Systems Division, NASA Lewis Research
Center, 21000 Brookpark Drive, Cleveland, Ohio 44135 as the person
to contact for the information desired by Mr. Clendenon.

Sincerely,

George P. Bates
GEORGE P. BATES, JR.
Director, Aircraft Development Service, DS-1

R. D. 1 COLUMBIA, N. J. 07832
PHONE (201) 496-4366

May 4th, 1970

Mr. William Clendenon
P.O. Box 49
Port Angeles, Washington, 98362

Dear Bill:

    Haven't run into a good solid Irish name in a long
time! Correct: I have been working on this fascinating
business of mercury engines for quite some time now but,
since ever-more material comes in, I have felt that it would
be most inappropriate to write up anything on it yet. Ser-
ious-minded matters such as this take a very long time to
investigate and then, even if one does come up with something
worthwhile, it takes even a cheap magazine about eight months
to get it into print --even if they like it, approve it, and
it fits into their schedule.

    I have your material on file, and in due course you
will be hearing from me further. I haven't the foggiest
notion what Argosy or any other magazine might want to do
with any story that I wrote on this subject. If any publication should wish to do so, and I even mentioned your
work, I (personally) would, of course, credit you fully; and,
should anything get published, I would seek your further
permission in advance and endeavour to come to some business
arrangement with you. This, Mr. Clendenon, is exceedingly
interesting work on your behalf and I hope only that it will
one day get published in a form that the ordinary citizen
can understand.

                   Yours very sincerely,

                   Ivan (T. Sanderson)

ITS:as
Enc:1
                  Ivan T. Sanderson.

159

DEPARTMENT OF THE AIR FORCE
HEADQUARTERS UNITED STATES AIR FORCE
WASHINGTON, D.C.

2 September 1970

Mr. Bill Clendenon
P. O. Box 49
Port Angeles, Washington 98362

Dear Mr. Clendenon

Your recent letter addressed to the Secretary of Defense,
concerning VTOL Gas Turbine Aircraft (Electromagnetic), has
been referred to me for reply.

Your designs have been thoroughly reviewed by technical
personnel in Air Force research and development. Although
the designs were found to be interesting, I am sorry to
advise you that they cannot be included in our development
program.

Thank you for your letter and the time you have taken to
bring this to our attention.

I am returning your photostats herewith.

Sincerely

M. A. MADSEN
Lt Col, USAF
Special Assistant
DCS/Research & Development

1 Atch
Photostats (7)

DIRECTIONAL GYRO

flew along the path defined by a space stabilized directional gyro, it would be flying straight in space. This could not be the same as flying North, since flying North is a curved line in space. This effect is sometimes called apparent drift due to meridian convergence It is also called *North steaming error*—and the amount of error depends apon speed and latitude.

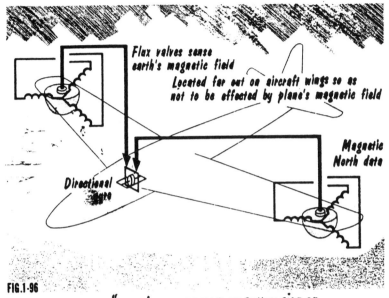

Flux valves sense
earth's magnetic field
Located far out on aircraft wings so as
not to be effected by plane's magnetic field

Magnetic
North data

Directional
gyro

FIG.1-96

NOTE: "TWO PING PONG BALLS ON BOTTOM SIDE OF
AIRPLANE WING". BILL CLENDENON U.F.O. RESEARCH.

Flux Valve

The directional gyro has the same random drift problems common to the vertical gyro. Even originally pointed North, random torques due to pickoff friction, mechanical unbalance, etc., would cause the gyro to precess in an unpredictable fashion.

Therefore we must have a control system which would keep the spin axis of the directional gyro always pointing North. Why not use a compass as a reference, as we used the pendulum for the vertical gyro? Basically, that is what is done, except that it is difficult to hook up a position pickoff to a compass needle. The forces available are so small that most any measuring device would foul up compass performance. Instead, we hang a flux valve (see Fig. 1-96) in the earth's magnetic field. A flux valve is much like a syncrho. In the synchro, a particular set of voltages is set up in the three stator winding—depending upon the orientation of the rotor with

NOTE: PAGE - (1-91) - FROM TEXT BOOK.

COMPONENTS

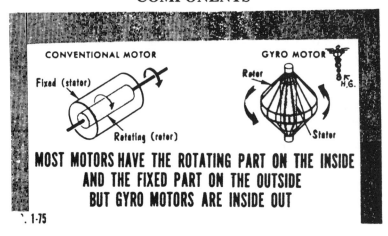

CONVENTIONAL MOTOR   GYRO MOTOR

Fixed (stator)   Rotor

Rotating (rotor)   Stator

MOST MOTORS HAVE THE ROTATING PART ON THE INSIDE AND THE FIXED PART ON THE OUTSIDE BUT GYRO MOTORS ARE INSIDE OUT

. 1-75

A rotor and stator assume that the gyro motor is actually an electric motor, and in most instances this is the case. See Fig. 1-75. However, many aircraft gyros used in panel instruments employ an air jet to spin the rotor. In some missile applications, where the rotor must reach full speed in just a few seconds, the driving force may be a clockspring or an explosive cartridge. Where this type of drive is used, the gyro may only be needed for a few minutes.

Wheels driven as electric motors usually operate between 4,000 and 24,000 rev./min. Air driven rotors may spin as rapidly as 100,000 rev/min.

Note **Switch and Electromagnetic Pendulous Reference Devices**

In a bang-bang erection system a mercury switch is often used as a pendulous reference. These switches are housed in a glass tube and can be built to carry high power. Erection torquers can be driven directly from the switches without the need for amplifiers. See Fig. 1-76.

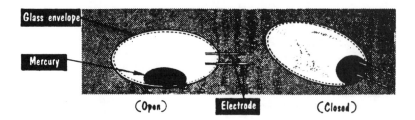

Glass envelope

Mercury

(Open)   Electrode   (Closed)

FIG. 1-76   MERCURY SWITCH MAY BE USED IN A BANG-BANG ERECTION SYSTEM

NOTE: LOOK WHATS INSIDE - (1-73) - THE PING PONG BALLS!

TEXT BOOK.   BILL CLENDENON U.F.O. RESEARCH.

The excellent research efforts of Leslie & Cramp I found very helpful in my own research program. But to carry things a bit further, the general dimensions as described in Fig. 15b below can be gone into further. According to my own observations of a vehicle as Cramp's drawing describes, I will have to disagree that any part of the power plant, etc., extends below the base of the outer flange of the ship, for the following reasons: (1) The outer flange must be stationary to act as a support for the whole ship while at rest on any surface. (2) The power plant has to be clear in order to revolve or run, thus the outer flange, while acting as a support overall, also allows running clearance for the turbines, etc. Some of the other uses for the outer flange include acting as a plenum chamber, an air plane, dissipating of heat, etc. In regard to the so-called landing gear or spheres, according to some of my research findings, it is within possibility that they may act as an emergency landing-gear by acting as a hydraulic shock absorber in some cases, though this might interfere with the running of the power plant, etc. They could also act as floats or buoyancy chambers in the case of a water landing. But the rest of the dimensions as described in Fig. 15b seem nearly correct according to my own observation of a similar type vehicle. I think Leslie & Cramp both deserve a great deal of credit and recognition for their research efforts. NOTE: To anyone, official or otherwise, who claims to have or had absolute proof in any shape or form, that Adamski's Scout Ship photo was faked in any manner, I say this: Why didn't you produce your absolute proof while Adamski was alive, and why don't you do it now before Dr. Condon, etc., make known the findings of their UFO investigations? Or is your absolute evidence not so absolute and you're just playing it safe? If you've had absolute proof that would've cleared up the Adamski case all this time and have not made it known, you have done an injustice to the public & this brings up the question of your motives. *William D. Clendenon Jr.* 5/28/67

Information compiled & arranged by WDCjr (including notations).

Obtain glossy print photocopies for better evaluation.

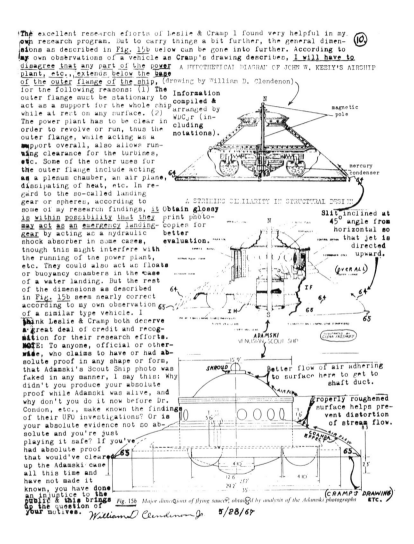

A HYPOTHETICAL DIAGRAM OF JOHN W. KEELY'S AIRSHIP (drawing by William D. Clendenon)

magnetic pole

mercury condenser

A STRIKING SIMILARITY IN STRUCTURAL DESIGN

Slit inclined at 45° angle from horizontal so that jet is directed upward.

(OVERALL)

ADAMSKI VENUSIAN SCOUT SHIP

Better flow of air adhering to surface here to get to shaft duct.

Properly roughened surface helps prevent distortion of stream flow.

COANDA EFFECT AIR FLOW

SHROUD

Fig. 15b *Major dimensions of flying saucer, obtained by analysis of the Adamski photographs*

(CRAMP'S DRAWING ETC.)

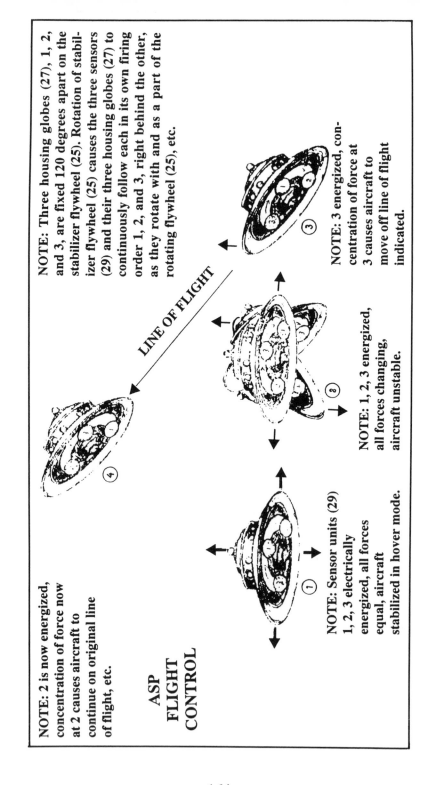

NOTE: Three housing globes (27), 1, 2, and 3, are fixed 120 degrees apart on the stabilizer flywheel (25). Rotation of stabilizer flywheel (25) causes the three sensors (29) and their three housing globes (27) to continuously follow each in its own firing order 1, 2, and 3, right behind the other, as they rotate with and as a part of the rotating flywheel (25), etc.

LINE OF FLIGHT

NOTE: 3 energized, concentration of force at 3 causes aircraft to move off line of flight indicated.

③

②

NOTE: 1, 2, 3 energized, all forces changing, aircraft unstable.

④

①

NOTE: Sensor units (29) 1, 2, 3 electrically energized, all forces equal, aircraft stabilized in hover mode.

NOTE: 2 is now energized, concentration of force now at 2 causes aircraft to continue on original line of flight, etc.

ASP
FLIGHT
CONTROL

164

ASP
FLIGHT
CONTROL

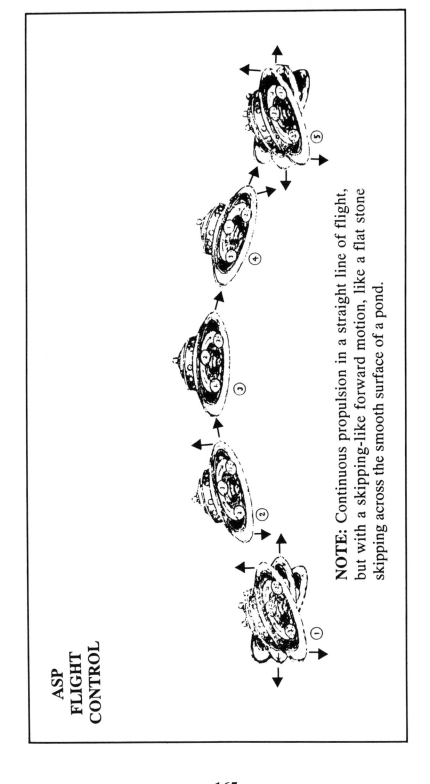

NOTE: Continuous propulsion in a straight line of flight, but with a skipping-like forward motion, like a flat stone skipping across the smooth surface of a pond.

165

UFO RESEARCH

To Develop Practical Military, Space, and Research Applications of

ELECTROHYDRODYNAMICS

As Related To:

SPACE PROPULSION............................Electrical

ENERGY CONVERSION.......................Electrical to Mechanical

FLUID PUMPING...............................No Moving Parts

HARD VACUUM PUMP......................10^{-9} mm. of Hg or Better

CONTINUOUS PARTICLE GUN............100,000 fps meteorite simulator

ELECTRIC POWER GENERATOR.........Flame-Jet Generator

Submitted by: Prepared by:

WHITEHALL-RAND, INC.   SEABROOK HULL

1019 Dupont Circle Bldg.   AND ASSOCIATES

Washington 6, D. C.   March 4, 190

1957-1960 - T.T. Brown hired as a consultant for
Whitehall-Rand Project under Auspices of Bahnson Labs
(Agnew Bahnson) to do antigravity R. & D.

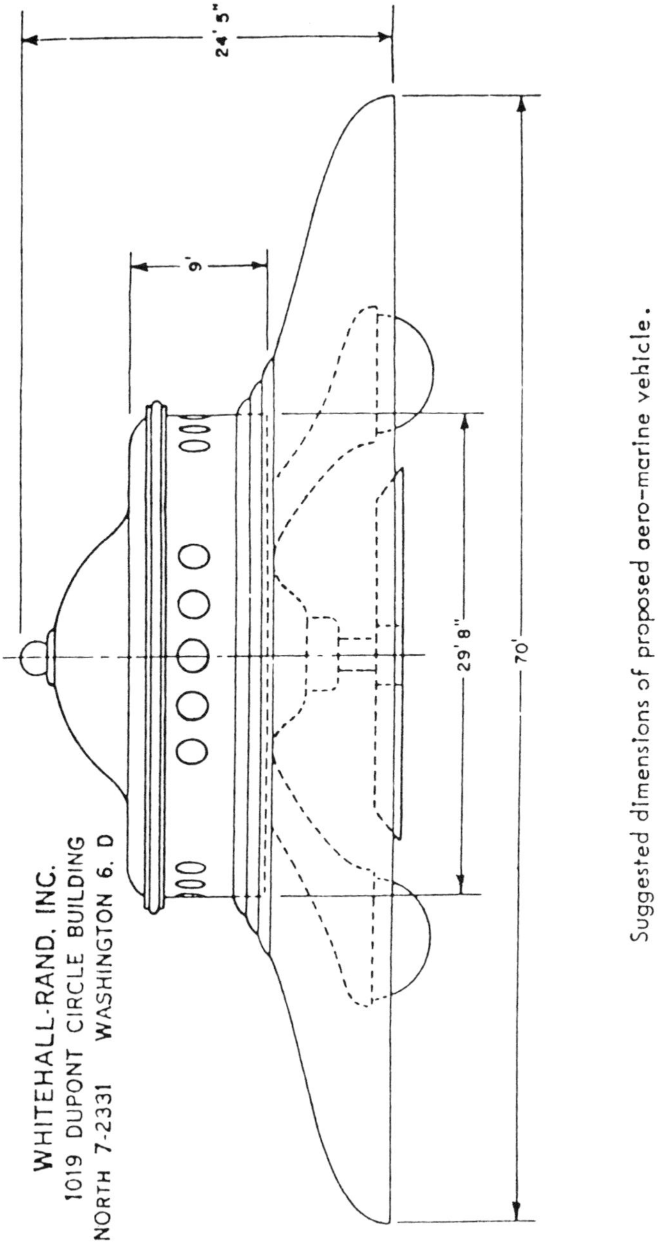

24' 5"

9'

29' 8"

70'

Suggested dimensions of proposed aero-marine vehicle.

WHITEHALL-RAND. INC.
1019 DUPONT CIRCLE BUILDING
NORTH 7-2331   WASHINGTON 6. D

1957-1960 - T.T. Brown hired as a consultant for
Whitehall-Rand Project under Auspices of Bahnson Labs
(Agnew Bahnson) to do antigravity R. & D.

167

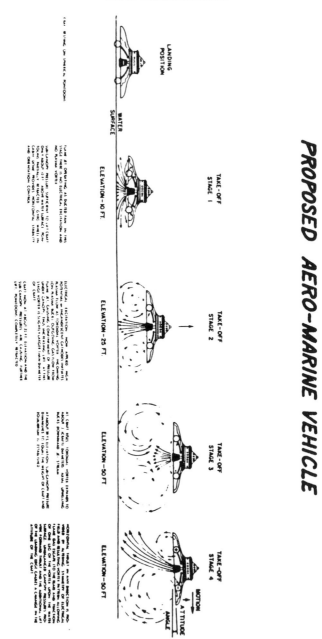

1957-1960 - T.T. Brown hired as a consultant for
Whitehall-Rand Project under Auspices of Bahnson Labs
(Agnew Bahnson) to do antigravity R. & D.

AFTERWORD

by
David Hatcher Childress

Arthur C. Clarke's Laws:

1) When a distinguished, but elderly scientist states that something is possible, he is almost certainly right. When he states that something is impossible, he is very probably wrong.

2) The only way to discover the limits of the possible is to look beyond them into the impossible.

3) Any sufficiently advanced technology is indistinguishable from magic.

The Rama Empire of India

Unlike other ancient nations such as Egypt, China, Brittany or Peru, the ancient Hindus did not have their history books all ordered destroyed, and therefore we have one of the few true links to an extremely ancient and scientifically advanced past. Modern scholars value ancient Hindu texts, as they are one of the last tenuous connections to the ancient libraries of the past. The super-civilization known as the Rama Empire is described in the *Ramayana*, which holds many keys to the truth of the past.

The *Ramayana* describes the adventures of a young prince named Rama who marries a beautiful woman named Sita. After some years of marriage, Sita runs off with (or is kidnapped by) Rama's enemy, Ravanna. Ravanna takes Sita by vimana to his capital city on an island called Lanka. Rama uses his own vimana, and a small army of friends, to

171

fly to Lanka and get his troublesome wife back. He brings her back to his home city, Ayodhya, where she banishes herself to the forest for being unfaithful. Rama, after years of anguish, finally reunites with her, and they live happily ever after.

> *Rama ruled the earth for 11,000 years.*
> *He gave a year-long festival*
> *In this very Naimisha Forest.*
> *All of this land was his kingdom then;*
> *One age of the world ago;*
> *Long before now, and far in the past.*
> *Rama was king from the center of the world,*
> *To the Four Oceans' shores.*
> —the beginning chapter of the
> *Ramayana* by Valmiki

The Vimanas of Ancient India

Nearly every Hindu and Buddhist in the world—hundreds of millions of people—has heard of the ancient flying machines referred to in the *Ramayana* and other texts as vimanas. Vimanas are mentioned even today in standard Indian literature and media reports.

In 1875, the *Vimanika Sastra*, a fourth-century BC text written by Maharishi Bhardwaj, was rediscovered in a temple in India. The book (taken from older texts, says the author) dealt with the operation of ancient vimanas and included information on steering, precautions for long flights, protection of the airships from storms and lightning, and how to switch the drive to solar energy, or some other "free energy" source, possibly some sort of "gravity drive." Vimanas were said to take off vertically, and were capable of hovering in the sky, like a modern helicopter or

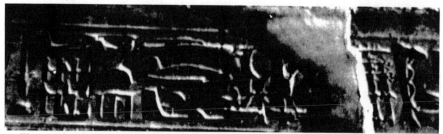

The symbols at Abydos Temple in Egypt look identical to a modern helicopter, a rocket, a flying saucer-type craft and a jet. The Indian texts called them vimanas.

dirigible. Bhardwaj the Wise refers to no less than seventy authorities and ten experts on air travel in antiquity. These sources are now lost.[10, 33]

The ancient Indian epics go into considerable detail about aerial warfare over 10,000 years ago—so much detail that a famous Oxford professor included a chapter on the subject in a book on ancient warfare!

According to the Sanskrit scholar Ramachandra Dikshitar, the Oxford professor who wrote *War in Ancient India* in 1944, "No question can be more interesting in the present circumstances of the world than India's contribution to the science of aeronautics. There are numerous illustrations in our vast Puranic and epic literature to show how well and wonderfully the ancient Indians conquered the air. To glibly characterize everything found in this literature as imaginary and summarily dismiss it as unreal has been the practice of both Western and Eastern scholars until very recently. The very idea indeed was ridiculed and people went so far as to assert that it was physically impossible for man to use flying machines. But today what with balloons, aeroplanes and other flying machines, a great change has come over our ideas on the subject."[5]

Says Dr. Dikshitar, "...the flying vimana of Rama or Ravana was set down as but a dream of the mythographer till aeroplanes and zeppelins of the present century saw the light of day. The *mohanastra* or the "arrow of unconsciousness" of old was until

An Assyrian cylinder seal depiction of a winged disk.

173

An Assyrian cylinder seal depiction of three men in a winged disk.

very recently a creature of legend till we heard the other day of bombs discharging poisonous gases." 5

The ancient texts also made the important distinction that vimanas were real machines, while contact with the spirit world, angels or fairies was a different matter. Says Dikshitar:

> The ancient writers could certainly make a distinction between the mythical which they designated *daiva* and the actual aerial wars designated *manusa*. Some wars mentioned in ancient literature belong to the *daiva* form, as distinguished from the *manusa*. An example of the daiva form is the encounter between Sumbha and the goddess Durga. Sumbha was worsted and he fell headlong to the ground. Soon he recovered and flew up again and fought desperately until at last he fell dead on the ground. Again, in the famous battle between "the celestials" and the Asuras elaborately described in the *Harivamsa*, Maya flung stones, rocks and trees from above, though the main fight took place in the field below. The adoption of such tactics is also mentioned in the war between Arjuna and the Asura Nivatakavaca, and in that between Karna and the Raksasa in both of which, arrows, javelins, stones and other missiles were freely showered down from the aerial regions.

King Satrujit was presented by a Brahman Galava with a

horse named Kuvalaya which had the power of conveying him to any place on the earth. If this had any basis in fact it must have been a flying horse. There are numerous references both in the *Visnupurana* and the *Mahabharata* where Krishna is said to have navigated the air on the Garuda. Thither the accounts are imaginary or they are a reference to an eagle-shaped machine flying in the air. Subrahmanya used a peacock as his vehicle and Brahma a swan. Further, the Asura, Maya by name, is said to have owned an animated golden car with four strong wheels and having a circumference of 12,000 cubits, which possessed the wonderful power of flying at will to any place. It was equipped with various weapons and bore huge standards. ...After the great victory of Rama over Lanka, Vibhisana presented him with the Puspaka vimana which was furnished with windows, apartments, and excellent seats. It was capable of accommodating all the Vanaras besides Rama, Sita and Laksmana. Rama flew to his capital Ayodhya pointing to Sita from above the places of encampment, the town of Kiskindha and others on the way. Again Valmiki beautifully compares the city of Ayodhya to an aerial car.

This is an allusion to the use of flying machines as transport apart from their use in actual warfare. Again in the *Vikramaurvasiya,* we are told that king Puraravas rode in an aerial car to rescue Urvasi in pursuit of the Danava who was carrying her away. Similarly in the *Uttararamacarita* in the fight between Lava and Candraketu (Act VI) a number of aerial cars are mentioned as bearing celestial spectators. There is a statement in the *Harsacarita* of Yavanas being acquainted with aerial machines. The Tamil work *Jivakacintamani* refers to Jivaka flying through the air.[5]

Mercury Engines and Vimana Texts

Bill Clendenon was primarily interested in the information on mercury contained in the *Vimanika Shastra.*

In chapter five of the *Vimanika Shastra,* Bhardwaj describes from the ancient texts which are his reference, how to create a mercury vortex engine:

Prepare a square or circular base of 9 inches width with wood and glass, mark its centre, and from about an inch and a half thereof draw lines to edge in the 8 directions, fix 2 hinges in each of the lines in order to open shut. In the centre erect a 6 inch pivot and four tubes, made of *vishvodara* metal, equipped with hinges and bands of iron, copper, brass or lead, and attach to the pegs in the lines in the several directions. The whole is to be covered.

Prepare a mirror of perfect finish and fix it to the *danda* or pivot. At the base of the pivot an electric *yantra* should be fixed. Crystal and glass beads should be fixed at the base, middle, and end of the pivot or by its side. The circular or goblet shaped mirror for attracting solar rays should be fixed at the foot of the pivot. To the west of it the image-reflector should be placed. Its operation is as follows:

First the pivot or pole should be stretched by moving the *keelee* or switch. The observation mirror should be fixed at its base. A vessel with mercury should be fixed at its bottom. In it a crystal bead with hole should be placed. Through the hole in the chemically purified bead, sensitive wires should be

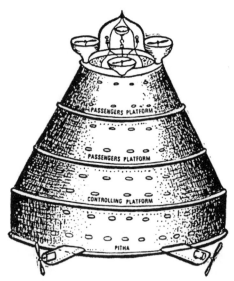

According to the Vimana texts, there were 4 kinds of vimanas.
1. The Rukma Vimana, a disc or circular craft.
2. The Sundara Vimana, also circular and pointed, like a rocket.
3. The Shakuna Vimana, a winged craft with a central tower.
4. The Tripura Vimana, a tubular or cigar-shaped craft.

passed and attached to the end beads in various directions. At the middle of the pole, a mustard cleaned solar mirror should be fixed. At the foot of the pole a vessel should be placed with liquid *ruchaka* salt. A crystal should be fixed in it with hinge and wiring. In the bottom centre should be placed a goblet-like circular mirror for attracting solar rays. To the west of it a reflecting mechanism should be placed. To the east of the liquid salt vessel, the electric generator should be placed and the wiring of the crystal attached to it. The current from both the *yantras* should be passed to the crystal in the liquid *ruchaka* salt vessel. Eight parts of sun-power in the solar reflector and 12 parts of electric power should be passed through the crystal into the mercury and on to the universal reflecting mirror. And that mirror should be focused in the direction of the region which has to be photographed. The image which appears in the facing lens will then be reflected through the crystal in the liquid salt solution. The picture which will appear in the mirror will be true to life, and enable the pilot to realize the conditions of the concerned region, and he can take appropriate action to ward off danger and inflict damage on the enemy.[3]

Two paragraphs later Bhardwaj says:

Two circular rods made of magnetic metal and copper should be fixed on the glass ball so as to cause friction when they revolve. To the west of it a globular ball made of *vaatapaa* glass with a wide open mouth should be fixed. Then a vessel made of *shaktipaa* glass, narrow at bottom, round in the middle, with narrow neck, and open mouth with 5 beaks should be fixed. Then a vessel made of *shaktipaa* glass, narrow at bottom, round in the middle, with narrow neck, and open mouth with 5 beaks should be fixed on the middle bolt. Similarly on the end bolt should be placed a vessel of sulfuric acid (*bhraajaswad-draavada*). On the pegs on the southern side 3 interlocked wheels should be fixed. On the north side a liquefied mixture of load-stone, mercury, mica, and serpent-slough should be placed. And crystals should be placed at the requisite centres.

'Maniratnaakara' [here Bhardwaj is referring to an ancient authority, now lost—ed.] says that the *shaktyaakarshana* yantra should be equipped with 6 crystals known as *Bhaaradwaaja, Sanjanika, Sourrya, Pingalaka, Shaktipanjaraka,* and *Pancha-jyotirgarbha.*

The same work mentions where the crystals are to be located. The *sourrya mani* is to be placed in the vessel at the foot of the central pole. *Bhaaradwaaja mani* should be fixed at the foot of the central pole. *Sanjanika mani* should be fixed at the middle of the triangular wall. *Pingalaka mani* is to be fixed in the opening in the *naala-danda. Pancha-jyotirgarbha mani* should be fixed in the sulfuric acid vessel, and *Shakti-panjaraka mani* should be placed in the mixture of magnet, mercury, mica, and serpent-slough. All the five crystals should be equipped with wires passing through glass tubes.

Wires should be passed from the centre in all directions. Then the triple wheels should be set in revolving motion, which will cause the two glass balls inside the glass case, to turn with increasing speed rubbing each other the resulting friction generating a 100 degree power...[3]

From the text of the *Vimaanika Shastra* it is apparent that mercury, copper, magnets, electricity, crystals, gyros(?) and other pivots, plus antennas, are all part of at least one kind of vimana. The recent resurgence in the esoteric and scientific use of crystals is interesting in the context of the *Vimaanika Shastra.* Crystals (*mani* in Sanskrit), are apparently as integral a part of vimanas as they are today of a digital watch. It is interesting to note here that the familiar Tibetan prayer Om Mani Padme Hum, is an invocation to the "Crystal (or jewel) inside the Lotus (of the mind)."

While crystals are no doubt wondrous and important technological tools, it is mercury that concerns us here.

Mercury is an element and a metal. It is a metallic element that was known to the ancient Chinese, Hindus, and Egyptians. The chief source of mercury is cinnabar HgS, a mineral. According to *Van Nostrand's Scientific Encyclopedia,* Mercury was mined as early as 500 BC out of cinnabar crystals which are usually "small and often highly modified hexagonal

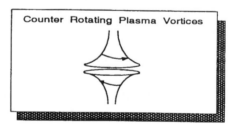

Counter Rotating Plasma Vortices

crystals, usually of rhombohedral or tabular habit. Its name is supposed to be of Hindu origin."

Mercury was most certainly mined and used earlier than 500 BC; scientific encyclopedias and such are usually overly conservative. The metal was named after the messenger of the gods in Roman mythology. It is a heavy, silver-white liquid with the symbol *Hg*. The symbol for Mercury is derived from the Greek word *Hydrargos* meaning water, silver or liquid gyro. It is a liquid at ordinary temperatures and expands and contracts evenly when heated or cooled.

The liquid metal mercury, when heated by any means, gives forth a hot vapor that is deadly. Mercury is generally confined to glass tubes or containers that are sealed, and therefore harmless to the user. Present day mercury vapor turbine engines use large quantities of mercury, but little is required for renewal because of its closed circuit systems. Mercury and its vapor conduct electricity; its vapor is also a source of heat for power usage. Mercury amplifies sound waves and doesn't lose timbre in quality. Ultrasonics can be used for dispersing a metallic catalyst such as mercury in a reaction vessel or a boiler. High-frequency sound waves produce bubbles in liquid mercury. When the frequency of the bubbles grow to match that of the sound waves the bubbles implode, releasing a sudden burst of heat.

Clendenon's concept is of a mercury-filled flywheel that can be used for stabilization and propulsion in discoid aircraft/spacecraft. Liquid mercury proton gyroscopes, according to Clendenon, can be used as direction-sensing gyros if placed 120 degrees apart on the rotating stabilizer flywheel of a discoid craft.

Liquid mercury proton gyroscopes have several

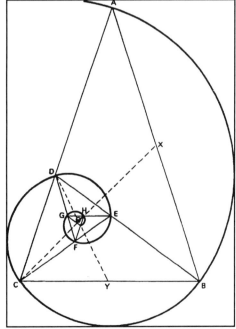

Above: Logarithmic spiral with embedded phi proportioned triangles. From *Quest For Zero-Point Energy* by Moray B. King.

advantages, says Clendenon. Firstly, the heavy protons found in mercury atoms are the very stable. Secondly, such gyros do not require a warm-up period as mechanical gyros do. Thirdly, the gyro using stable mercury protons is not affected by vibrations and shock. Fourthly, the liquid mercury proton gyroscope has no moving parts and can run forever. And lastly, the mercury atom offers the most stable gyro device in nature and has the additional advantages of saving space and weight. This is particularly valuable on long distance flights where all space and weight must be very carefully calculated and conserved.

Clendenon has done a great deal of experimentation with mercury vortex technology in the context of the ancient texts. His vimana, modeled after Adamski's "scout ship," consists of a circular air frame that is partly a powerful electromagnet through which is passed a rapidly pulsating direct current. It works basically like this:

- The electromagnetic field coil, which consists of the closed circuit heat exchanger/condenser coil circuit containing the liquid metal mercury and/or its hot vapor, is placed with its core axis vertical to the craft.

- A ring conductor (directional gyro-armature) is placed around the field coil (heat exchanger) windings so that the core of the vertical heat exchanger coils protrudes through the center of the ring conductor.

- When the electromagnet (heat exchanger coils) is energized, the ring conductor is instantly shot into the air, taking the craft as a complete unit along with it.

- If the current is controlled by a computerized resistance (rheostat), the ring conductor armature and craft can be made to hover or float in the Earth's atmosphere.

- The electromagnet hums and the armature ring (or torus) becomes quite hot. In fact, if the electrical current is high enough, the ring will glow dull red or rust orange with heat.

- The phenomenon (outward sign of a working law of nature) is brought about by an induced current effect identical with an ordinary transformer.

- As the repulsion between the electromagnet and the ring conductor is mutual, one can imagine the craft being affected and responding to the repulsion phenomenon as a complete unit.

- Lift or repulsion is generated because of close proximity of the field magnet to the ring conductor. Clendenon says that lift would always be vertically opposed to the gravitational pull of the planet Earth, but repulsion can also be employed to cause fore and aft propulsion.

Clendenon thus interprets the *Samaran Sutradhara* quite differently than most scholars, and voila—"By means of the power latent in the mercury which sets the driving whirlwind in motion a man sitting inside may travel a great distance in the sky in a most marvelous manner."[105]

Clendenon's interpretation of much of the discoid craft seen since 1947 is that many are vimanas, either of ancient manufacture, or modern manufacture. He believes that the famous scout ship observed by George Adamski (and later by other witnesses) is neither a hoax nor an interplanetary space craft. His mercury vortex engines are not capable of interplanetary flight, he says, but, like this version of a vimana, are for terrestrial flight only. He believes that a great number of UFO phenomena could be explained as effects of mercury vortex technology, and craft using it. He thought that some of these craft were ancient—flown by mysterious humans who lived for hundred of years—and that some of them were modern constructions, made by the Americans, British, and Germans.

As to unusual UFO effects, he says that the ball of light that often surrounds the UFO ship is the magneto-hydrodynamic plasma, a hot, continuously recirculating air flow through the ship's gas turbine which is ionized (electrically conducting). According to Clendenon, at times a shimmering mirage-like effect caused by heat, accompanied by pulsations of the ball of light, makes the craft appear to be alive and breathing. This has, at times, suggests Clendenon, made witnesses to certain UFOs think that they were seeing a living thing. For some of the above reasons, the ship may seem to suddenly disappear from view, though it is actually still there and not de-materialized. The ionized bubble of air surrounding the UFO may be controlled by a computerized rheostat so the ionization of the air may shift through every color of the spectrum, obscuring the aircraft from view.

Curiously, the following item appeared on the Internet in 1998 concerning the U.S. government's secret aircraft called the TR-3B, which was claimed to be powered by a mercury vortex drive as described in the *Vimanika Shastra*:

"The TR-3B Triangular Anti-Gravity Craft, by Ed Fouche: A very important speech was given by Ed Fouche to the *1998 Summer Sessions at the International UFO Congress*, describing the 200 foot across triangular UFO 'anti-gravity' craft being built and tested in area S-4 inside Area 51 in Nevada. Supposedly uses a heated mercury vortex to offset gravity 'mass.'"

> *The wheels sparkled like topaz and they were all alike:*
> *in form and working they were like a wheel inside a wheel,*
> *and when they moved in any of the four directions*
> *they never swerved course.*
> —Ezekiel, 1:16-17

Advanced Forms of Electric Generation

Clendenon's ideas on mercury-plasma-gyros follow along with advances in electric generation. In the last 50 years the power industry has begun to examine other possibilities of electric generation other than three-phase dynamos. Among these new devices, rarely seen in mainstream literature are Magnetohydrodynamics, Electro-gas-dynamics, and Thermoelectric and Thermionic Generators. All of these advanced devices figure into Clendenon's so-called Mercury Proton Gyros.

In magnetohydrodynamics (MHD), an ionized gas or liquid metal is passed through a magnetic field to generate electricity. Experimental units have been constructed both in the United States and elsewhere, but none is yet capable of being substituted for conventional generators.

Electro-gas-dynamics (EGD) utilizes a gas stream to carry charged particles through an electric field. The electric field opposes the motion of the particles and slows them down, thereby increasing their charge by converting their kinetic (motion) energy into direct current.

Clendenon's ideas on mercury-plasma-gyros follow along with advances in electric generation. In the last 50 years the power industry has begun to examine other possibilities of electric generation other than three-phase dynamos. Among these new devices, rarely seen in mainstream literature are Magnetohydrodynamics, Electro-gas-dynamics, and Thermoelectric and Thermionic Generators. All of these advanced devices figure into Clendenon's so-called Mercury Proton Gyros.

In magnetohydrodynamics (MHD), an ionized gas or liquid

metal is passed through a magnetic field to generate electricity. Experimental units have been constructed both in the United States and elsewhere, but none is yet capable of being substituted for conventional generators.

Electro-gas-dynamics (EGD) utilizes a gas stream to carry charged particles through an electric field. The electric field opposes the motion of the particles and slows them down, thereby increasing their charge by converting their kinetic (motion) energy into direct current.

As has been done with MHD, some encouraging results have been obtained with EGD but many problems remain to be solved before EGD can replace conventional generating-plant equipment.

Thermoelectric and Thermionic Generators are generators working on the principle of thermoelectricity or of thermionic emis-

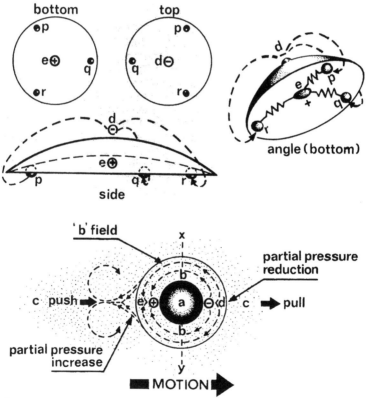

Plasma Saucer in Motion. Stan Deyo's model of a mercury proton-type discoid craft in motion. From *The Cosmic Conspiracy* by Stan Deyo.

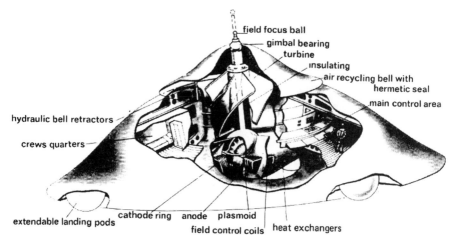

Plasma Saucer in Motion. Stan Deyo's model of a mercury proton-type discoid craft in motion. From *The Cosmic Conspiracy* by Stan Deyo.

sion are static devices that can convert heat directly into electricity without converting it into mechanical energy first. Both have been used in some small applications with low power requirements, but neither seems promising as a source of large amounts of utility power in the near future.

Another recent development in energy that may effect is the power supply for Clendenon's Mercury Proton Gyros is Fuel Cells. Fuel cells are thought by some to have a bright future as sources of residential and industrial electric power. The fuel cell is an electro-chemical device that converts the chemical energy of the fuel directly into a direct-current electrical output, somewhat like a continuous-process battery. Hydrogen-fueled cells have been used in space applications, and a group of natural-gas utilities has sponsored an extensive research program to develop a natural-gas fuel cell, but so far none is competitive with conventional utility-electric service.

Many physicists and authors have suggested plasma gyro-type craft similar to Clendenon's, such as Moray B. King and Stan Deyo.

Deyo popularized the plasma gyro saucer in the fine technical illustrations in his underground bestseller *The Cosmic Conspiracy*. Plasma Saucer in Motion. Stan Deyo's model of a mercury proton-type disoid craft in motion. Deyo claimed in his book that many

UFO sightings were of secret discoid craft being manufactured by the U.S. Military (in conjuction with a super-secret group of British and Australian intelligence) in an underground base somewhere in the center of Australia. Such places as Pine Gap near Alice Springs (featured in the film *The Falcon and the Snowman*) or the Woomera military installation further south, have been suggested. Deyo's underground factory for these plasma saucers may be in totally secret location that is connected to the partially underground facilities of Pine Gap and Woomera by underground tunnels—maglev trains, ofcourse.

Deyo claims in his book that he actually worked on these craft and describes them in detail. Like Clendenon, he claims that electrical-conducting gases, such as the liquid metal Mercury, are moved in a vortex-gyro pattern and used to lift various "experimental" flying saucers. Such a craft would fly as a discoid hovercraft, and, because of the powerful electrified whirling metal particles, would have bright lights as blinding as the largest neon sign ever built.

In my efforts to find out the viability of these electrified gas-plasma craft, I contacted physicist-engineer Moray B. King, and asked him what he thought of Clendenon's concept of an electrified mercury gyroscope that gave an anti-gravity lift effect. He replied that it was a sound principle, one worth studying.

In his new book *Quest Zero Point Energy*[30] King includes several chapters on a number of amazing technologies that could come out of electrified mercury gas (plasma) being forced into a gyroscopic vortex.

Says King:

The primary hypothesis for tapping the ZPE is that stimulating a plasma into a self-organized coherent form induces a like coherence in the virtual plasma of the quantum foam. This would suggest that a glow plasma be subjected to one or more of the following stimulations to induce a coherent nonlinear self-organization:
1. Abrupt EM pulse
2. Bucking EM fields
3. Counter-rotating EM fields
There are "free energy" inventions associated with each

stimulation. For example... all utilize the abrupt discharge in their inventions to manifest a plasmoid form as well as excess energy. Correa, [T. Henry] Moray and Papp apply the discharge directly to a glow plasma. Shoulders pulses a liquid metal electrode and Graneau pulses a small cylinder of water. Shoulders, Correa and Graneau have observed plasmoid formations via photographic methods. Shoulders, Correa and Moray tapped the excess energy via rectifying an output pulse, while Papp and Graneau have focused on tapping the anomalously large mechanical reaction force via a piston. It is noteworthy that an abrupt discharge in a glow plasma or liquid also produces a characteristic bucking field compression. Mesyats has described in detail the behavior of a liquid surface in response to an electric discharge. It forms a stalk protruding from the surface, which is symmetrically surrounded by a polarized glow plasma. The tip

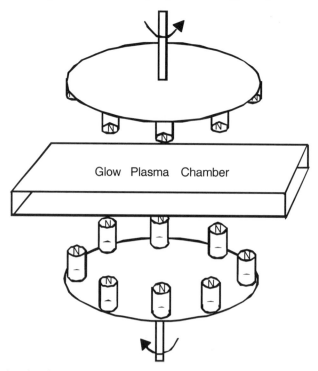

Stimulating the glowing mercury plasma can be done with counter rotating, bucking magnetic fields as well as an abrupt electric discharge from a pulsing circuit (not shown). From *Quest For Zero-Point Energy* by Moray B. King.

of the stalk explodes into the glow plasma yielding a perfectly symmetrical compression by two ion layers. It is the symmetry that guides the emitted electron plasma to form into a closed helical toroidal filament, producing the EV or plasmoid.

A number of inventors have used radioactive materials to help create the glow plasma. Moray is perhaps the most famous where his "Swedish stone" cathode contained a mixture of luminescent and radioactive compounds pressed into a germanium pellet to make a "radioactive transistor" with sufficient gain to drive a small loudspeaker. Papp also used radium and luminescent materials in his electrodes to help ionize his inert gas mixture. Brown used this principle in a simplified fashion to make his nuclear battery where a weak radioactive source creates a glow plasma which interacts resonantly with an LC circuit to produce anomalously excessive energy output. The simplicity of Brown's invention makes it a good candidate for replication.

The use of bucking EM fields to produce a scalar excitation in the vacuum energy has been emphasized by Bearden. Abruptly opposing EM fields produce a stress on the fabric of space and increase the electric potential, yet since the fields are in opposition, they sum into a net zero field vector. Nonetheless the abrupt stress and release can cause an orthorotation of the ZPE flux, and can couple vacuum energy into the glow plasma being so stimulated. Caduceus coils or Mobius coils have been suggested for such excitation. Recent Soviet experiments with Mobius coils have claimed to launch "non-orientable entities" akin to ball lightning, as well as produce negative energy formations described as "magnetic monopoles." Subjecting a glow plasma to abruptly bucking magnetic fields might produce some large ZPE effects.

The Power System of the Gods

And so we see that modern physics is coming full circle with the ancient vimana texts and Bill Clendenon's ideas. Is mercury the element of the gods? Is the caduceus is a virtual diagram for a mercury vortex propulsion device? The late William Clendenon thought so. In the next section, Part Two, join us as we explore the world of Tesla Technology and the ancient world of Atlantis. This book has been thousands of years in the making, and hopefully it will be worth the wait...

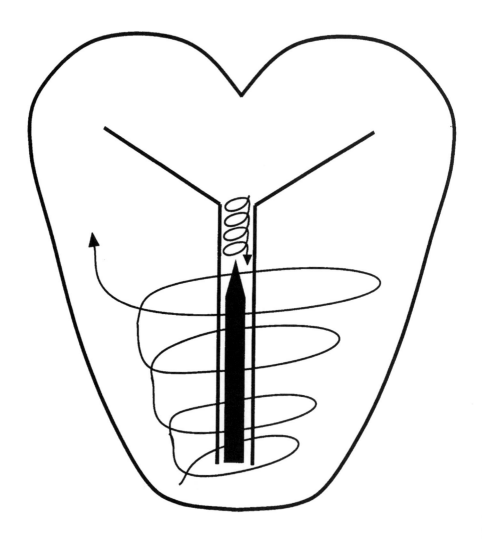

Cross section of a vortex globe designed to make the "anu" shaped, dual vortex. Funnel guides flow to the inner pipe where a tapered rod induces a spiral. A magnetic stirrer (not shown) can act as a propeller to control the spin in a sealed globe. Some fine sand mixed into an air vortex would electrostatically charge, and its circulation might mimic a plasma vortex. If the vortex induced a large scale quasi charge in the zero-point energy, the globe would exhibit a positive or negative polarity depending on the direction of spin. From *Quest For Zero-Point Energy* by Moray B. King.

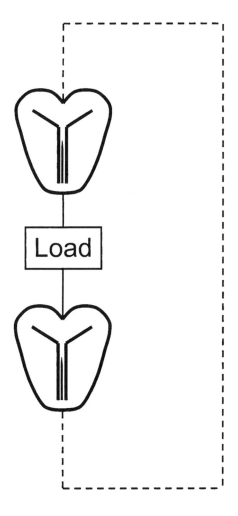

Experiments with two "anu" vortex globes spinning oppositely might exhibit cold current where thin wires guide vortical, vacuum polarization displacement current. If the vortex globes produce a pair of large quasi charges, which act like a flux source and sink, a complete circuit loop (dashed line) might not be required. From *Quest For Zero-Point Energy* by Moray B. King.

PART TWO

THE TESLA– ATLANTEAN POWER SYSTEM

BY

DAVID HATCHER CHILDRESS

PART 2
CHAPTER 1

THE AMAZING WORLD OF
ATLANTEAN TECHNOLOGY

Galileo's Conclusion:

*Science proceeds more by what it has learned to ignore
than what it takes into account.*

Arthur C. Clarke's Law:

It has yet to be proven that intelligence has any survival value.

The Strange Case of Colonel Percy Fawcett

In 1893, a young British officer named Percy Fawcett was stationed
in Ceylon. Operating out of Trincomalee and keenly interested in
archaeology, history and Buddhism, he would often take long walks,
sometimes lasting for days, into the remote jungle areas of the island.
On one such trek, he was overtaken by a storm, which forced him to
seek shelter beneath some trees for the night. As dawn broke into a
new, sunny day, he found himself near
an immense rock, covered with strange
inscriptions of unknown character and
meaning.

He made a copy of the inscriptions,
and later showed them to a local Buddhist priest. This priest said the writing was similar to that used by the old
Asoka-Buddhists, and was in a cipher
which only those ancient priests could

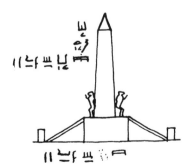

Colonel Percy Fawcett

understand. His assertion was confirmed ten years later by a Ceylonese Oriental scholar at Oxford University, who claimed that he was the only man alive who could read the script.[34, 39]

Young Percy Fawcett, later to become a respected colonel and one of the most famous South American explorers of all time, believed that the letters which he had seen on the ancient, vine-covered wall in Ceylon had been taken from the ancient Sansar alphabet. This alphabet was first discovered by the French traveler and missionary, Abbe Huc, in 1845 while visiting a lamasery on the frontier of Tibet and China. The lamasery, known as the monastery of Sinfau or Sifau, or more popularly, the Kumbum Monastery, contained the "mystic Kounboum tree"; upon each leaf of this tree a Sansar character was allegedly written. *Kunbum* or *Kounboun* means "ten thousand images" referring to the images on the leaves of the tree.

According to the report given to Huc, the tree and alphabet came from the drowned land of Rutas, which in central Asian mythology is a lost civilization now beneath the ocean. It is often identified with "Atlantis" although it could be identified with "Mu" or Lemuria, or some other unidentified civilization now beneath the Indian Ocean, the Indonesian Sea or off the coast of China and Japan.

The name of *Rutas* for a lost continent showed up again in 1879, this time by another French traveler named Louis Jacolliot (1837-1890). Jacolliot was a serious student of mythology, and had collected a number of Sanskrit legends while on sojourn in India. His book, *Histoire des Vierges: Les Peuple et les Continents Disparus*, a study of mythology, was published in Paris in 1879. According to Jacolliot, the Hindu classics tell of a former continent called *Rutas* that sank beneath the ocean in times past, according to the traditions of the *goparams*. Jacolliot believed Rutas to have been a former Pacific continent, that the original inhabitants of India had come from this vanished continent and that the language of *Rutas* was Sansar.[39]

A number of events were set into motion by young Percy Fawcett's finding of these strange characters on a rock wall in Ceylon, some involving mysteries still to be worked out. Colonel Fawcett set off from Cuyaba in the Mato Grosso in Brazil in 1925 to find a lost city in the jungle. He believed there was a connection between the city and the letters he had found in Ceylon. Neither he nor his two companions were ever seen again, and their expedition became the archetypical "lost expedition."

Expedition Fawcett To a Lost World

Percy Harrison Fawcett was quite successful in his army career, leading eight South American expeditions under contract with the Bolivian and Brazilian governments, to delimit the frontiers shared by these two countries with Peru and Ecuador. Between the years of 1906 and 1922, he had made four arduous journeys in Bolivia and three in Brazil, as well as other expeditions into Peru and Ecuador.

At a lecture before the Royal Geographical Society in London in 1911, Fawcett described a "lost world" on the borders of Bolivia and Brazil, and told of gigantic footprints of primeval monsters he had seen. Sir Arthur Conan Doyle, the creator of Sherlock Holmes, was present at the lecture, later writing a book based on Fawcett's tales, *The Lost World: The Adventures of Professor Challenger.* You may have seen the movie version, made in the 1950s, on late night TV.

Later, H. Rider Haggard, author of *King Solomon's Mines,* gave Fawcett a mysterious stone idol. Haggard had allegedly received it from the British Consul, O'Sullivan Beare, who had picked it up at a lost city in Brazil in 1913. This stone statue was in the possession of Colonel Fawcett when he disappeared in the jungle in 1925, but its story didn't end there.

Fawcett, a believer in the paranormal, had several "sensitives" examine the statuette, in order to ascertain its origin. He wrote that they believed that the idol came from Atlantis. Fawcett himself was a great believer in Atlantis, and felt that the lost cities in the interior of Brazil had an Atlantean origin. He disagreed with one popular theory for the origin of Atlantis, which actually placed the lost civilization in Brazil, but believed that Brazil was once a colony of Atlantis.

Fawcett also believed that some of the writing copied at the end of the old Portuguese document was identical to the writing he had seen years earlier in Ceylon, and that both writings originated in Atlantis. He hoped to prove the existence of Atlantis by rediscovering this lost city.

As will usually happen, several of Fawcett's assumptions were erroneous. Despite the psychics' opinions to the contrary, the idol originated in the Mediterranean area around 400 BC. According to Barry Fell, an author who has deciphered many ancient inscriptions, the foot-tall, basalt idol was made at or around Hallicarnassus before Hellenistic times. It is an image of a priest of Baal, advertising his temple, dedicated to Hercules (Melgart, son of Baal and Tanitte). The language is Creole Minoan-Hittite, according to Fell. It read, in two parts, "To ask the Gods for a lucky omen of the future, invoke Melgart and... bring a propitiation for him."

In the ancient Mediterranean, an area of many diverse cultures, it was common for different countries to form alliances and work together for economic or political ventures. Therefore, it is not unusual that a combination of languages such as Minoan-Hittite would be used on a statuette left at one of the lost cities.

Colonel Fawcett reported just before his ill-fated expedition that he had been told that a Nafaqua Indian chief, whose territory lay between the Xingu and Tabatinga Rivers, claimed to know of a city where strange temples could be found, and baptismal ceremonies were practiced. The Indians there spoke of houses with "stars to light them, which never went out."

Said Fawcett in his book, "This was the first but not the last time I heard of these permanent lights found occasionally in the ancient houses built by that forgotten civilization of old. I knew that certain Indians of Ecuador were reputed to light their huts at night by means of luminous plants, but that, I considered, must be a different thing all together. There was some secret means of illumination known to the ancients that remains to be rediscovered by the scientists of today—some method of harnessing forces unknown to Us."[23, 12]

Brian Fawcett then added a footnote to his father's book: "In view of recent developments in atomic research there is no reason to dismiss the 'lamps that never go out' as myth. The world was plunged into a state of barbarism by terrible cataclysms. Continents subsided into the oceans, and others emerged. Peoples were destroyed and the few survivors who escaped were able to exist only in a state of savagery. The ancient arts were all but forgotten, and it is not for us in our ignorance to say that the science of antediluvial days had not advanced beyond the level we have now

Colonel Percy Fawcett and companion find giant reptile tracks in the Amazon.

reached."

There is today a method of growing a quartz crystal with phosphorous dispersed throughout its interior. Such a crystal will absorb daylight and then emit that light at night. This would be a simple, yet ingenious device for creating a light that shines by itself, a light storage battery which could sit on the top of a pyramid or pillar for years, shining every night! Can small quartz crystal devices such as decorative skulls or pyramids absorb radiated power from an obelisk or Tesla Tower and then later glow all night? Fawcett was a believer in all of these things.

On May 29, 1925, Colonel Fawcett wrote a letter to his wife, Nina, from Dead Horse Camp, deep in the Mato Grosso. Fawcett's horse had died on the same spot in 1920, forcing him to turn back. He told Nina, "You need have no fear of any failure..."

These were the last words ever written by the Colonel, and his disappearance became an enduring mystery, especially in Britain, Brazil and Peru, where the Fawcett family had lived and worked over many years. Starting in 1928, expeditions of various sorts took off to the jungles of Brazil in search of Fawcett and his companions. One expedition even had Peter Fleming, the brother of novelist Ian Fleming of James Bond fame, as a member. In 1933 a report claimed that an "English Colonel" was being held prisoner by a remote tribe and in the following year a Royal Geographic Society expedition discovered a theodolite compass belonging to Fawcett.

The Fate of Colonel Fawcett

In 1951 a Portuguese book published an account of a 1943 expedition which included a confession by a Kalapalos Indian chief, Izariri. Izariri allegedly claimed that Fawcett and his two companions, his oldest son, Jack and another young man named Raleigh Rimmel, had been killed by the tribe. That same year, senor Orlando Vilas Boas of the Central Brazil foundation published a confession by chief Izarari, that he had clubbed the two Fawcetts and Rimell to death. Izariri's successor, Comatzi, disclosed the alleged grave of Colonel Fawcett, where bones were subsequently dug up and sent to England for examination. After a team of experts from the Royal Anthropological Institute in London examined these remains, they declared that they could not be the bones of Colonel Fawcett. It is possible that they may have been those of Albert de Winton, however, who was lost searching for Colonel Fawcett in 1930.

From this evidence, much of which is undoubtably idle storytelling, it

Colonel Percy Fawcett and companions at Dead Horse Camp.

seems possible that Colonel Fawcett, Jack and Raleigh were still alive as late as 1935, an incredible ten years after they had begun their expedition. But Brian Fawcett, in the final chapter of his father's book, draws negative conclusions from the evidence, not believing that his brother fathered a child with an Indian girl, and questioning the story brought back by Rattin of the nameless English Colonel. Quoting Brian, "And why, why did the old man not tell his name?"[23]

Actually, there is a good reason why the real Colonel may have remained silent. Brian himself tells the reader in the prologue to *Exploration Fawcett* that his father, before setting out on this last journey, "...fearful of other lives being lost on his account, urged us to do everything possible to discourage rescue expeditions should his party fail to come back."[23] And there may have been other reasons Fawcett chose to not disclose his identity. He may actually have preferred to stay with the Indians, though certainly his son and wife could not have imagined this.

Colonel Fawcett was a believer in psychic phenomena, as was his friend, Sir Arthur Conan Doyle. In 1955, an interesting book called *The Fate of Colonel Fawcett*[23] was published by the Aquarian Press in London. This rare book is an investigation into the disappearance of Colonel Fawcett by psychic Geraldine Cummins, who allegedly "makes contact" with the Colonel on several successive instances. As unconventional as its topic may be, the book makes fascinating and exciting reading. It reads, curiously enough, like an H. Rider Haggard novel, full of mystery, lost cities, savages, and evil priestesses.

According to *The Fate of Colonel Fawcett*, the Colonel was still alive in 1935, when the "contact" first started. Raleigh and Jack were killed by the Indian tribe that held them captive, when they insisted on continuing on to the lost city that they sought. Jack Fawcett was well liked by the Kalapalos, but the Indians felt that Rimell was devious, influencing Jack against them. As the two set off on their quest, the Indians who were supposedly escorting them shot both Jack and Rimell on the orders of chief Izariri. Jack was killed instantly with a dozen arrows in his back; Rimell was allowed to suffer for a few hours, as he was deemed the instigator in wanting to take Jack Fawcett away from the tribe.

The reason Izariri wanted to keep the explorers captive was that the chief had lived for some time among Europeans, and did not want the Whites' civilization to affect his tribe. He was afraid that if he let Fawcett's party loose, they would return with more Europeans. Izariri wanted the whites to believe that Colonel Fawcett was dead, and even provided proof so that they would discontinue the search. This could

The statue discovered in Brazil and given to Fawcett.

be why Albert de Winton was killed, and his remains then passed off as those of Colonel Fawcett.

According to *The Fate of Colonel Fawcett*, the Colonel did eventually reach the lost city, after the death of Jack and Raleigh. Izariri had wanted Colonel Fawcett to marry his sister, a high priestess. This woman hated Fawcett, and vice versa. The union never took place, as before the marriage, Fawcett insisted he visit the lost city, finally being escorted there safely by an Indian servant. Fawcett eventually died back in the village of the Kalapalos, poisoned by the priestess. So ended the bizarre story of Colonel Fawcett and his ill-fated expedition, as related through the medium Geraldine Cummins.

In the most bizarre part of this book, Fawcett says in the sup-

THE FATE OF COLONEL FAWCETT

BY

GERALDINE CUMMINS

AQUARIAN PRESS

The 1955 book *The Fate Colonel Fawcett.*

posed communication that he could imagine Egyptians walking in the city and that there were towers everywhere. Cummins has some curious information from "Fawcett" concerning the possible towers of Atlantis. In bits and pieces Fawcett, from the "other side," tells Cummins what it is like in Atlantis. In many of the readings references are made to towers as well as to Ireland, England, Atlantis, the lost land of Hy Brasil, Egypt, and other ancient lands. Fawcett describes himself at one point as walking among these towers, which used electricity from the atmosphere, and that the towers were an important part of Atlantis.

In a transcript from January 16, 1949, we learn that "I was in a valley, gazing up at a strange white town. I could vaguely perceive gigantic buildings, soaring towers. I was intrigued and thrilled and strove to project myself towards that distant vision; the effort only seemed to increase the distance between it and me. I tried again with all my might to go forward, and was suddenly faced by a great door of gleaming metal."

Are these towers some kind of obelisks, part of a Tesla-type system also described by Edgar Cayce and the Lemurian Fellowship, as we will see in later chapters? In Fawcett's astral travels in search of his lost city he believes that Atlantis lies beyond this door of gleaming metal—a door he must open. Alas, he does not open it, but is confronted by a giant snake, to him the symbol of death. Fawcett's is a strange story, one that, via Geraldine Cummins and her friends, brings us a small step closer to assembling a picture of the ancient power system of Atlantis (for more information on the fascinating story of Colonel Fawcett and his lost city, see my book, *Lost Cities & Ancient Mysteries of South America*).

The Bizarre Towers of Atlantis According to Colonel Fawcett

The strange world of ancient Atlantis/Egypt/South America is described by an entity who identifies himself as Colonel Fawcett to a

198

British spiritualist group on December 5, 1935, with Geraldine Cummins, her "control," a man named Astor, and others present including E. B. Gibbes, who was writing down the words coming from Ms. Cummins.

After some preliminary discussions the discarnate Colonel Fawcett begins to describe the "ancient city" that he is wandering in:

> I see the Indian girl and the chief occasionally still, but I don't know whether they are real or whether this old Egyptian world is real—that is my problem at present. Which condition is life? Which dream?

> *E.B.G. [E. B. Gibbes]: I asked him if he thought it possible that he was Passing over into the Beyond and dreaming of the city he wanted to find.*

> I rather imagine you are right in suggesting that I am passing over, that I am, in part, in one world, in part in another. It is an extraordinary experience. I wish I could get back to London and tell the people there of it. The priestess, the Lotus girl, is beckoning to me now.

> *The next sitting took place two days later, on December 7th, 1935, and after the last few words of the above writing had, at the alleged Colonel Fawcett's request, been read over to him, the following continuation of the record was written:*

> When I set out on that expedition to find the pyramids, a number of people said, and many thought, that I was crazy. But I am nothing of the sort. If we are to continue, you must believe in my sanity. You must accept my assurance that the last relics of an ancient civilisation, Egyptian in character, are to be found in central South America. With my living eyes I have seen these ruins. When still physically fit, I walked over that haunted country. It is haunted, you know, and unlike any other part of the earth for this reason. Can you imagine a heavy heat, rather breathless, great swamps, huge forests, luxurious growth everywhere? I believe that, if the climate were not so oppressive and we could bring gangs of men here, excavating under skilled direction, a whole ancient civilisation would be revealed—the secret of the Lost Continent would be divulged,

a flood of light thrown on a period that is pre-historic, and our origins more clearly realised. What is more, these races, were as civilised as are the Europeans of today. Only they travelled on a different orientation.

Sun worship was the basic principle on which the whole of this central American civilisation was founded. Now this is important.

Listen! You don't know, no living man knows, what electricity is. These Atlanteans knew more or less the nature of electricity, which is dependent on the sun yet is also allied to other air forces. Of course, there is more than one kind of electricity. The kind that is known to men was discovered by these Atlanteans, but they used their kind of electricity in a different way from us. They realised that it might be used, not merely to give light—queer globular lights—but that it might also be employed in connection with the shifting of weights. The building of the pyramids is solved when you know that huge blocks of stone can be manipulated through what I might call blast-electricity. You will think me mad when I talk of electrified winds, for you know nothing about the connection between air and electricity, the alliance between it and light. Terms like the compression of air and the accumulation of electricity for the purpose of combining the two so that what is solid may be removed, have not yet entered into the imagination of man. But I, who have seen this ancient world, walked through its streets, halted before the porticoes of its temples, descended into the great subterranean world wherein electricity and air are combined and fused, can assure you that the men who came before modem history was recorded, knew more about matter and light, about the ether and its properties, than the scientists of the twentieth century can ever know or imagine.

Picture to yourself wide reservoirs of compressed electrified air reclaimed and stored in huge pockets under the surface of the earth. Coal mines—Oh, yes, I know all about the miles of burrows under Welsh soil. But these are nothing when compared with the storage batteries that were like some vast design existing under the solid crust of soil, and were guarded and maintained by an army of Atlanteans. You won't find traces of these reservoirs, of these great store-houses of power, under the surface of the earth of central South America. For everything caved in in that respect, and the whole land was readjusted through the cataclysm. That is why the present ruins are but a very small fragment of what existed in those early days.

But this is what I am getting at. It was man, and not the forces of nature, that destroyed Atlantis. Or rather, men developed to such a degree these subterranean storehouses of electrified air, that at last it revolted and pitched man and the solid earth heavenward. Then great was the fall. Seas flowed in over the riven, sunken land. Thousands of miles of country were submerged and earth thrown up in other places formed new countries.

I am describing all this in a very crude way. But I think you will agree with me that a force that can move the stone necessary for the building of the pyramids as in Egypt, is of no mean order, and when with discovery and further discovery, this force is increased a millionfold in power, then the danger point has been reached.

I don't think your scientists have the imagination to conceive the principles of *electrified air—and I hope they never will. For you must understand that it can be used as a destructive weapon— extremely dangerous because it is invisible.

*According to the book's author, Geraldine Cummins, the above script was written two years before the atom was split in 1937, and ten years before men first learnt of the destructive powers of the atom bomb.

The book continues with the supposed comments of the discarnate Colonel Fawcett who is wandering the astral plane: "Men believed that they could become gods through acquiring knowledge of the secrets of the sun, light and colour. They became in their power almost like gods, and thus the myth of the fallen angels was created. Who are the fallen angels? They are an historic fact, and came out of the long-lost Atlantean story. You see, these men could control nature through their knowledge of electricity, of light; and the knowledge led to the Fall—to the destruction of Atlantis. Now, perhaps, you will realise the source of the old Biblical story, which is wrapped up in symbol but, for all that, it is not myth but fact."[24]

The chapter continues with the Spiritualists taking a break and then returning to their questioning of Colonel Fawcett:

Some hours after the conclusion of the above sitting, Astor came again and wrote: "I think your friend from South America can be easily summoned. Fawcett is intrigued by this method of communication, and he is in a curious state following still in the track of his dream. He is caught in the web of that dream and his mind is still bent on reaching its heart, on finding the source of life. For his ambition has increased as he has learned that sun-power is indeed the dynamo from which all material things on this earth derive the force necessary for movement and action."

The communicator [Cummins] *then took control and wrote what follows:*

I am tuning into your rhythm. I see so clearly now the nature of environment and how we hitch on to each one.

Here the communicator drew across the page three long wavy lines, of varying design, one under the other.

That is the story of the universe and of communicating intelligences, of communicating appearances.

Rhythms, uncountable rhythms, and we can only illustrate them by these undulating lines. We, on earth, live in one rhythm. But our soul is so constituted that if we choose to exercise its powers to the full, we can tune into other rhythms— enter again the various phases in the history of the earth, for mind and memory pervade the whole universe. We have but to adjust ourselves to another key and the melody of the Roman period, or the time of Charlemagne, sounds for us, and all the images of such a time pass before us, our perceptions registering them on certain occasions almost as clearly as sight and touch register the material facts of life for man on earth. But there are conditions attached to the entry of any one rhythm.

As a spectator is caught in the illusion of some remarkable drama and so temporarily forgets his own small history—forgets, for instance, the London grumbling past the doors of the theatre in which he is seated, so are we snatched away temporarily from what seemed essentially to belong to us—memories. of facts and details connected with our past story. We lose, indeed, a grip of our own immediate history when, for instance, we study that ancient Atlantean past and enter into its rhythm.

I cannot tell you now to which world I belong. I don't know if I still live in that little hut in an Indian village, or whether I am what is called dead. For I am back at times in that village, I speak with those Indians, and I lie in that hut. But this experience has a dream-like quality for me now. Reality, indeed, seems to reside for me away from that world, and in another which belongs to the Atlantean period of time.

I spoke to you before of the sun worship and the lost wisdom gleaned from this religion of the creative material life. Well, I have been closely studying it since I last spoke to you. For to me it is the finest adventure upon which I have entered so far. I want to get at the secret of the sun-power. I have noted, for instance, the many white towers—funicular in shape—that arise out of, and near to, these ancient cities of South America. They are all registered there as on a photographic plate. I note that these towers must be in some way conductors of electricity. And I have come to the conclusion (1) that they are connected with the vast subterranean reservoirs of power, (2) that this ancient people had learnt how to draw the electricity out of the atmosphere by means of these towers, (3) that the atmosphere of the earth is—and was even more pronouncedly so— very fully charged with electricity.

The scientists of modern times are correct in supposing that the sun is bombarding the earth with electrons, and that these are distributed by air currents.

These towers I have mentioned might be called sieves. I mean by that, that through a certain process they suck in the air currents and eject all but the electrons necessary for electric power. We on earth know how to generate electricity from coal and from water-power. These ancients went much further. They had devised an instrument that could extract from the all-pervading atmosphere the electricity necessary to all the concerns of their life. They used those massed electrons for a thousand different purposes. These not merely heated and lighted their dwellings, they moved great weights. Instruments were devised which automatically performed functions such as cooking, serving and cleaning in households. Furthermore, they were used for carrying, for healing, for defence.

The guns were invisible in those days. As the sun bombards

the earth with electrons, so was this race able to bombard its enemies with electrified blasts of air. They could wage war from a considerable distance. And this led to their undoing. More and more they accumulated electricity in their storehouses beneath the earth. They had to do it gradually, for they learnt that they might not wholly eliminate the electric particles from the atmosphere; for if they attempted to do so difficulties arose in connection with the health of the people.

However, eventually they were able to conserve this massed sun-power to such an extent that, when war broke out, they expended too violently and suddenly what I might call their electrified projectiles. The chambers of compressed electricity were suddenly rent asunder, a vast cataclysm followed, and— as I have previously described to you—the face of the world was changed by these convulsions.

You must understand that this concentration of sun-power was mostly in the country now forming the Atlantic's bed. In the region of South America I was but on the fringe of that ancient world.

Egypt should not be called part of Atlantis. It seems that, in the second era, after the cataclysm, those I shall call the lesser Atlanteans made attempts to colonise other parts of the earth. Egypt was, for a short time, such a colony; in fact the most typical Atlantean strain lived in Egypt. In the other regions of the world inferior races survived. So, only in Egypt and on islands near it was some fragment of Atlantean wisdom preserved. But the Egyptians inherited a wholesome fear of the Sun God: after all, his depredations were very considerable.

There is a great deal more on the fate of Colonel Fawcett in the book, but this is the part that concerns us here. It was many years ago that I was first introduced to this curious text. While it may not be verified in any way, and the fate of Colonel Fawcett may never be known, it is still a fascinating and, I think, instructive discourse.

What Colonel Fawcett seems to be relating to the British spiritualists is a type of Atlantean power system that uses towers and which might be related to the wireless electrical power system that Nikola Tesla was attempting to implement from 1910 to 1925. In the next chapter we will examine the system of wireless power proposed by Tesla.

A 1908 painting by the well-known artist Mikalojus Ciurlionis entitled *Piramidziu Sonata. Allegro*. Ciurlionis' painting is on the Atlantis theme and includes pyramids and towers.

PART 2
CHAPTER 2

TESLA'S WIRELESS POWER DISTRIBUTION SYSTEM

Featherston's Accurate Steps to Systems Development:

1) Wild enthusiasm.
2) Disillustionment.
3) Total confusion.
4) Search for the guilty.
5) Punishment of the innocent
6) Promotion of nonparticipants.

The Serbian-American electrical engineer and inventor Nikola Tesla, was born in Smijlan, Croatia (then part of Austria-Hungary), on July 9, 1856, the son of a clergyman and an inventive mother. It was Tesla who devised the alternating-current systems that underlie the modern electrical power industry.

Tesla had an extraordinary memory, one that made learning six languages easy for him. He entered the Polytechnic School at Gratz, where for four years he studied mathematics, physics and mechanics, confounding more than one professor by an amazing understanding of electricity, an infant science in those days, that was far greater than theirs.

Nikola Tesla

His practical career started in 1881 in Budapest, Hungary, where he made his first electrical invention, a telephone repeater (the ordinary loudspeaker) and conceived the idea of a rotating magnetic field, which later made him world famous in its form as the modern induction motor. The polyphase induction motor is what provides power to virtually every industrial application, from conveyer belts to winches to machine tools

Tesla's mental abilities require some mention, since not only did he have a photographic memory, but he was able to use creative visualization with an uncanny and practical intensity. He describes in his autobiography how he was able to visualize a particular apparatus and was then able to actually test-run the apparatus, disassemble it and check for proper action and wear! During the manufacturing phase of his inventions, he would work with all blueprints and specifications in his head. The invention invariably assembled together without redesign and worked perfectly. Tesla slept one to two hours a day and worked continuously on his inventions and theories without benefit of ordinary relaxation or vacations. He could judge the dimension of an object to a hundredth of an inch and perform difficult computations in his head without benefit of slide rule or mathematical tables. Far from an ivory tower intellectual, he was very much aware of the issues in the world around him, and made it a point to render his ideas accessible to the general public by frequent contributions to the popular press, and to his field by numerous lectures and scientific papers.

Tesla immigrated to the USA in 1884, bringing with him the various models of the first induction motors. He worked briefly for Thomas Alva Edison, who as the advocate of direct current became Tesla's unsuccessful rival in electric-power development. In 1888, Tesla showed how a magnetic field could be made to rotate if two coils at right angles were supplied with alternating currents 90 degrees out of phase with each other. Tesla's induction motors were eventually shown to George Westinghouse. Westinghouse bought rights to the patents on this motor and made it the basis for the Westinghouse power system. It was in the Westinghouse shops that the induction motor was perfected. Numerous patents were taken out on this prime invention, all under Tesla's name.

During his short tenure working under Thomas Edison, Tesla created many improvements on Edison's DC motors and generators, but left under a cloud of controversy after Edison refused to live up to bonus and royalty commitments. This was the beginning of a rivalry which was to have ugly consequences later when Edison and his backers did everything in their power to stop the development and installation of Tesla's far more efficient and practical AC current delivery system and urban power grid. Edison put together a traveling road show which attempted to portray AC current as dangerous, even to the point of electrocuting animals both small (puppies) and large (in one case an elephant) in front of large audiences. As a result of this propaganda crusade, the state of New York adopted AC electrocution as its method of executing convicts. Tesla won the battle by the demonstration of AC current's safety and usefulness when his apparatus illuminated and powered the entire New York World's Fair of 1899.

Tesla wireless power transmitter being constructed at Wardenclyffe, NY.

It was Tesla's association and loose partnership with George Westinghouse that brought about the implementation of Tesla's amazing inventions, including the alternating current power we use today.

Westinghouse was born in Central Bridge, New York on Oct. 6, 1846, and was an American inventor and industrialist who during his lifetime obtained approximately 400 patents, including that on the air brake. In 1865 he patented a device for replacing derailed freight cars on tracks; three years later he developed the railroad frog, which permits the wheel on one rail of a track to cross another rail of an intersecting track. In 1869 he founded the Westinghouse Air Brake Company. By the early 1880s, Westinghouse had developed interlocking switches and a complete railroad signal system and established (1882) the Union Switch and Signal Company. In 1886 he founded the Westinghouse Electric Company, which established the use of alternating current for electrical generating and transmitting apparatuses and electrical appliances in the United States. Westinghouse died on March 12, 1914.

Without the guidance and business sense of Westinghouse, Tesla became increasingly ineffective in his ability to get his inventions funded and released to the public. Tesla's other inventions included the Tesla coil, a kind of transformer, and he did notable research on high-voltage electricity and wireless communication. He made little money from his work, however, and later lived as an eccentric recluse.

Tesla died in New York City on January 7, 1943, many of his dreams unrealized. On his death the FBI searched Tesla's apartment and seized various papers and technical diagrams that jhe had kept. What inventions were possibly seized by the FBI is not currently known.

The Fundamentals of Alternating Current

In order to understand the significance of Tesla's work and its relation to ancient science, we should first review some of the fundamentals of electrical theory. Essentially, there are two kinds of electrical current: A direct current (DC) always flows through a conductor in one direction, but an alternating current (AC) constantly reverses itself as a result of reversing electromotive force. One complete reversal is a cycle, and the number of cycles per second is the frequency of the alternating current. The standard frequency of alternating current in the United States and the rest of North America is 60 Hz (1 Hz, or hertz, equals 1 cycle per second); in Europe it is 50 Hz. Originally, only direct current was generated for public use. The enormous advantages of Tesla's alternating current were not realized until George Westinghouse developed the transformer in the late 19th century.

The transformer made it possible to change the voltage (and therefore the current) of AC by a simple, static device. This could not be done with DC. When electricity is transmitted, power loss is minimized by stepping up the voltage from the generator, thus transmitting a high voltage, and stepping down the voltage at the user's end. When required, DC is easily obtained from an AC current with a rectifier. Converting DC to AC requires an inverter, which is a more complex device. AC motors and alternators (AC generators) have greater reliability than their DC counterparts, because AC generators do not require commutators (metal sliprings for picking up current).

Buffalo, N.Y., was the first U.S. city to be lit using alternating current. With the large hydroelectric alternating current generators installed at Niagara Falls, New York, the area of upstate New York and Ontario became electrified with what is essentially the modern system of today. New York City became the first major city in the world to be electrified by modern AC power.

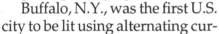
Tesla lights a special bulb from a wireless transmitter.

Theories on Electricity and Electric Charge

Electricity is a form of energy, a phenomenon that is a result of the existence of electrical charge. The theory of electricity and its inseparable effect, magnetism, is probably the most accurate and complete of all scientific theories. The understanding of electricity has led to the invention of motors, generators, telephones, radio and television, X-ray devices, computers, and nuclear energy systems. Probably no other discovery has transformed society and technology more than electricity. Yet, electricity was known to the ancients.

Amber is a yellowish, translucent mineral made from hardened (fossilized) tree sap. As early as 600 BC the Greeks were aware of its peculiar

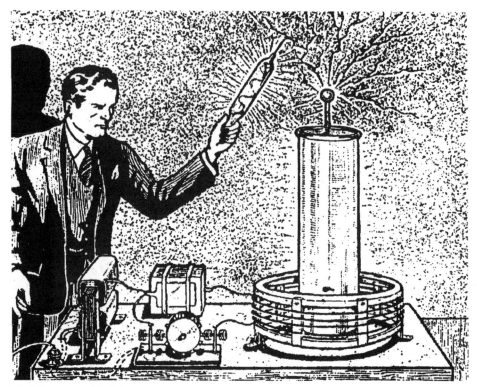

Tesla lights a special bulb from a wireless transmitter in this old drawing from *Electrical Experimenter*.

property: when rubbed with a piece of fur, amber develops the ability to attract small pieces of material such as feathers. Small sparks of electricity could be seen between the fur and amber in the dark. For centuries this strange, inexplicable property was thought to be unique to amber. The Greek word for amber is elektron, hence our modern word "electric."

Two thousand years later, in the 16th century, the English scientist William Gilbert proved that many other substances are electric and that they have two electrical effects. When rubbed with fur, amber acquires resinous electricity; glass, however, when rubbed with silk, acquires "vitreous" or static electricity. Electricity repels the same kind and attracts the opposite kind of electricity. Scientists thought that the friction actually created the electricity (their word for charge). They did not realize that an equal amount of opposite electricity remained on the fur or silk.

In 1747, Benjamin Franklin in America and William Watson (1715-87) in England independently reached the same conclusion: all materials possess a single kind of electrical "fluid" that can penetrate matter freely but that can be neither created nor destroyed. The action of rubbing merely transfers the fluid from one body to another, electrifying both. Franklin and Watson originated the principle of conservation of charge, which states that the total quantity of electricity in an insulated system is constant.

Feb. 1919 issue

Franklin defined the fluid, which corresponded to vitreous electricity, as positive and the lack of fluid as negative. Therefore, according to Franklin, the direction of flow was from positive to negative—the opposite of what is now known to be true. A subsequent two-fluid theory was developed, according to which samples of the same type attract, whereas those of opposite types repel.

Franklin was acquainted with the Leyden jar, a glass jar coated inside and outside with tinfoil. It was the first capacitor, a device used to store charge. The Leyden jar could be discharged by touching the inner and outer foil layers simultaneously, causing an electrical shock to a person. If a metal conductor was used, a spark could be seen and heard. Franklin wondered whether lightning and thunder were also a result of electrical discharge.

In 1752, during a thunderstorm, Franklin flew a kite that had a metal tip. At the end of the wet, conducting hemp line on which the kite flew he attached a metal key, to which he tied a nonconducting silk string that he held in his hand. The experiment was extremely hazardous, but the results were unmistakable: when he held his knuckles near the key, he could draw sparks from it.

It was known as early as 1600 that the attractive or repulsive forces diminishes as the charges are separated. This relationship was first placed on a numerically accurate, or quantitative, foundation by Joseph Priestley, a friend of Benjamin Franklin. In 1767, Priestley indirectly deduced that when the distance between two small, charged bodies is increased by some factor, the forces between the bodies is reduced by the square of the factor. For example, if the distance between charges is tripled, the force decreases to one-ninth its former value.

The French physicist Charles A. de Coulomb, whose name is used as the unit of electrical charge, later performed a series of experiments that added important details, as well as precision, to Priestley's proof. He also promoted the two-fluid theory of electrical charges, rejecting both the idea of the creation of electricity by friction and Franklin's single-fluid model.

Metal Conductors, Resistance, and the Speed of Electricity

Although a conductor permits the flow of charge, it is not without a cost in energy. The electrons are accelerated by the electric field. Before they move far, however, they collide with one of the atoms of the conductor, slowing them down or even reversing their direction. As a result, they lose energy to the atoms. This energy appears as heat, and the scattering is a resistance to the current.

In 1827 a German teacher named Georg Ohm demonstrated that the current in a wire increases in direct proportion to the voltage V and the cross-sectional areas of the wire A, and in inverse proportion to the length l. Because the current also depends on the particular material, Ohm's law is written in two steps, $I = V/R$, and $R = Pl/A$. The quantity R is called the resistance, and rho, which depends only on the type of material, is the resistivity. The unit of resistance is the ohm, where 1 ohm is equal to 1 volt/amp.

In lead, a fair conductor, the resistivity is $22 \times 10(-8)$ ohm-meters; in copper, an excellent conductor, it is only $1.7 \times 10(-8)$ ohm-meters. Where high resistances between 1 and 1 million ohms are needed, resistors are made of materials such as carbon, which has a resistivity of $1400 \times 10(-8)$ ohm-meters. Certain materials, such as lead, lose their resistance almost entirely when cooled to within a few degrees of absolute zero. Such materials are called superconductors. Substances have recently been found that become superconductive at much higher temperatures.

The resistive heating caused by electron scattering is a significant effect and is

Tesla demonstrating wirless power at his New York laboratory, 1899.

214

used in electric stoves and heaters as well as in incandescent light bulbs. In a resistor the power P, or energy per second, is given by $P = I^2R$.

As electrons bounce along through the wire, the general charge drift constitutes the current. The average, or drift, speed is defined as the speed the electrons would have if all were moving with constant velocity parallel to the field. The drift speed is actually small even in good conductors. In a 1.0-mm- diameter copper wire carrying a current of 10 amps at room temperature, the drift speed of the electrons is 0.2 mm per second. In copper, the electrons rarely drift faster than 10(-11) (one hundred-billionth) the speed of light. On the other hand, the speed of the electric signal is the speed of light. This means that, at the speed of light, the removal of one electron from one end of a long wire would affect electrons elsewhere.

Says *Groliers*, "For example, consider a long, motionless freight train, with the cars representing electrons in a wire. Because the couplings between cars have play in them, the caboose is affected a short while after the engine begins moving. During this time the engine moves forward a short distance. The signal telling the caboose to start moves backward quickly, traveling the length of the train in the same time it takes the engine to go forward a meter or so. Similarly, the electron drift speed in a conductor is

Tesla's wireless transmission tower in action, sending power to electrical airships in this illustration for *Radio News*, December, 1925.

low, but the signal moves at the speed of light in the opposite direction."

Quantum Theory and Unanswered Problems in Modern Electrical Theory

Early in the 20th century the quantum theory was developed. According to this theory, the electron is a smeared cloud of mass and charge. In some situations the electron cloud might be so small that the particle appears to be much like the tiny, charged marble of earlier views. In others, as when the electron is in an atomic orbit, the cloud is many times larger.

In 1963, Murray Gell-Mann and George Zweig of the California Institute of Technology proposed a theory according to which the electronic charge "e" might not be the fundamental charge after all. In their theory, heavy particles such as protons and neutrons consist of various combinations of particles called quarks. Quarks are supposed to have a charge of either - 1/3 e or + 2/3 e. The theory has since been expanded to include six types of quarks.

Despite the widespread success of electrical power, important unanswered questions remain within the field of electricity and electrical-atomic theory.

A basic question is: How does the force get from here to there? Perhaps it is by the exchange between charged particles of quanta of electromagnetic radiation. These hypothetical quanta are small, chargeless, massless particles in a so-called virtual state. This idea is part of the theory of quantum electrodynamics, developed by Richard Feynman of the California Institute of Technology and Julian Schwinger of Harvard. This theory is puzzling, however. The complete answer might never be known.

Another unsolved problem involves the electrical theory of matter. The electron is considered a small body packed with negative electrical charge. According to some scientists, it is a ball of charge having a radius of approximately 10(-15) meters. What holds it together? Unless some other force, an attractive one, is involved, the negative charge on one side repelling the negative charge on the other side would tear the particle apart. Another force may exist, although no such force has been found.

N. TESLA.

APPARATUS FOR TRANSMITTING ELECTRICAL ENERGY.

APPLICATION FILED JAN. 18, 1902. RENEWED MAY 4, 1907.

1,119,732.

Patented Dec. 1, 1914.

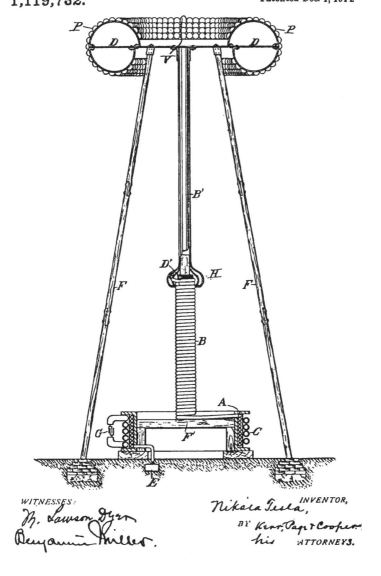

WITNESSES:
M. Lawson Dyer
Benjamin Miller.

INVENTOR,
Nikola Tesla,
BY Kerr, Page & Cooper
his   ATTORNEYS.

Tesla's 1914 patent on the broadcast tower for his wireless electricity.

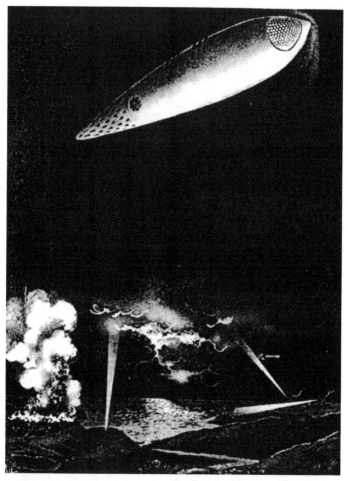

Tesla's concept of an anti-gravity airship with no wings or propeller, from the Oct. 1919 issue of *Electrical Experimenter.*

Modern Electric Power Transmission

Electric power has become an indispensable form of energy throughout much of the world. Even systems that use forms of energy other than electricity are likely to contain controls or equipment that run on electric power. For example, modern home heating systems may burn natural gas, oil, or coal, but most systems have combustion and temperature controls that require electricity in order to operate. Similarly, most industrial and manufacturing processes require electric power, and the computers and business machines of many offices and commercial establishments are paralyzed if electric service is interrupted.

During the first part of the 20th century, only about 10% of the total energy generated in the United States was converted to electricity. By 1990 electric power accounted for about 40% of the total. Developing countries are usually not as dependent on electricity as are the more industrialized nations, but the growth rate of electricity use in some of those countries is comparable to the rate of growth in the early years of electricity availability in the United States.

George Westinghouse

The first commercial electric-power installations in the United States were constructed in the latter part of the 19th century. The Rochester, N.Y., Electric Light Co. was established in 1880. In 1882, Thomas A. Edison's Pearl Street steam-electric station began operation in New York City and within a year was reported to have had 500 customers for the lighting services it supplied. A short time later a central station powered by a small waterwheel began operation in Appleton, Wisconsin.

In 1886 the feasibility of sending electric power greater distances from the point of generation by using alternating current (AC) was demonstrated at Great Barrington, Mass. The plant there utilized Westinghouse transformers to raise the voltage from the generators for a high-voltage transmission line.

The electric power industry of the United States grew from small beginnings such as these to become, in less than 100 years, the most heavily capitalized industry in the country. It now comprises about 3,100 different corporate entities, including systems of private investors, federal and other government bodies, and cooperative-user groups. Less than one-third of the corporate groups have their own generating facilities; the others are directly involved only in the transmission and distribution of electric power.

Electric power transmission systems today consist of step-up transformer stations to connect the lower-voltage power-generating equipment to the higher-voltage transmission facilities; high-voltage transmission lines and cables for transferring power from one point to another and pooling generation resources; switching stations, which serve as junction points for different transmission circuits; and step-down transformer stations that connect the transmission circuits to lower-voltage distribution systems or other user facilities. In addition to the transformers, these transmission substations contain circuit breakers and associated connection devices to switch equipment into and out of service, lightning arresters to protect the equipment, and other appurtenances for particular applications of electricity. Highly developed control systems, including sensitive devices for rapid detection of abnormalities and quick disconnection of faulty equipment, are an essential part of every installation in order to provide protection and safety for both the electrical equipment and the public.

Many of the first high-voltage transmission lines in the United States were built principally to transmit electrical energy from hydroelectric plants to distant industrial locations and population centers. High-voltage transmission lines were originally designed to permit the construction of large generating units and central stations on attractive, remote sites close to fuel sources and supplies of cooling water. Today, however, they connect different power networks in order to achieve greater economy by exchanges of low-cost power, to achieve savings in reserve generating capacity, to improve the reliability of the system, and to take advantage of diversity in the peak loads of different systems and thereby reduce operating costs.

At one time power lines in the 33-kV or 44-kV class were classified as high-voltage lines. As loads increased and transmission distances became greater, transmission voltages were increased. Electrical losses increase proportionately to the square of the current—the higher the voltage of the

Mark Twain and Joseph Jefferson in Tesla's New York laboratory, 1894, with a blurred image of Tesla between them.

ine, the lower the current needed to carry an equivalent amount of power. Moreover, one high-voltage line can usually carry as much power as several lower-voltage ones, so the use of higher voltages reduces the number of lines required and conserves the space required for rights-of-way. Voltage levels increased to 69, 115, 138, and 161 kV in various sections of the United States.

Tesla at his Colorado Springs laboratory with his "magnifying transmitter" capable of producing millions of volts of electricity.

Before World War II the highest-voltage lines in the United States were 230 kV, with the exception of one 287-kV line from Boulder Dam to Los Angeles. In the early 1950s several 345-kV lines were constructed. By 1964 the first 500-kV lines in the United States were being completed, and in 1969 the first 765-kV line was put into service. All of these involved AC systems.

In 1970 a 1,380-km (856-mi), 800-kV direct-current (DC) line was placed in commercial service to connect northwestern U.S. hydroelectric sources with the Los Angeles area. Such systems offer an economical means of transferring large quantities of power over long distances. They also avoid stability problems sometimes encountered by AC systems and DC systems are sometimes used to connect AC systems even over short transmission distances. Still, the transmission of power through wires has many limitations. Tesla decided that wireless transmission of power was the better way to go.

Tesla's System of Wireless Transmission of Power

Tesla's most important work at the end of the nineteenth century was his original system of transmission of energy by wireless antenna. In 1900 Tesla obtained his two fundamental patents on the transmission of true wireless energy covering both methods and apparatus and involving the use of four tuned circuits. In 1943, the Supreme Court of the United States granted full patent rights to Nikola Tesla for the invention of the radio, superseding and nullifying any prior claim by Marconi and others in regards to the "fundamental radio patent." It is interesting to note that Tesla, in 1898, described the transmission of not only the human voice, but images as well and later designed and patented devices that evolved into the power supplies that operate our present day TV picture tubes. The first primitive radar installations in 1934 were built following principles, mainly regarding frequency and power level, that were stated by Tesla in 1917.

In 1889 Tesla constructed an experimental station in Colorado Springs where he studied the characteristics of high frequency or radio frequency alternating currents. While there he developed a powerful radio transmitter of unique design and also a number of receivers "for individualizing and isolating the energy transmitted." He conducted experiments designed to establish the laws of radio propagation which are currently being "rediscovered" and verified amid some controversy in high energy quantum physics.

Tesla wrote in *Century Magazine* in 1900: "...that communication without wires to any point of the globe is practicable. My experiments showed that the air at the ordinary pressure became distinctly conducting, and this opened up the wonderful prospect of transmitting large amounts of electrical energy for industrial purposes to great distances without wires... its practical consummation would mean that energy would be available for the uses of man at any point of the globe. I can conceive of no technical

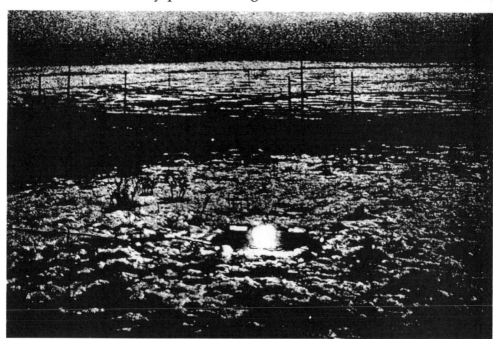

Tesla sends wireless power to three lights in one of his 1899 experiments at Colorado Springs.

advance which would tend to unite the various elements of humanity more effectively than this one, or of one which would more add to and more economize human energy..."

Over a century ago Tesla had devised a system of broadcasting electrical power through the atmosphere by utilizing a network of specially designed towers. This, I am arguing in this book, is essentially the same as the Atlantean power system in use many thousands of years ago.

After finishing preliminary testing, work was begun on a full-sized broadcasting station at Shoreham, Long Island. Had it gone into operation, it would have been able to provide usable amounts of electrical power at the receiving circuits. After construction of a generator building and a 180-foot broadcasting tower, financial support for the project was suddenly withdrawn by J. P. Morgan when it became apparent that such a worldwide power project couldn't be metered and charged for.

According to Toby Grotz of the (now defunct) International Tesla Society, it has been proven that electrical energy can be propagated around the world between the surface of the Earth and the ionosphere at extreme low frequencies in what is known as the Schumann Cavity. The Schumann Cavity surrounds the Earth at ground level and extends upward to a maximum 80 kilometers. Experiments to date have shown that electromagnetic waves of extreme low frequencies in the range of 8 Hz, the fundamental Schumann Resonance frequency, propagate with little attenuation around the planet within the Schumann Cavity. Knowing that a resonant cavity can be excited and that power can be delivered to that cavity similar to the methods used in microwave ovens for home use, it should be possible to resonate and deliver power via the Schumann Cavity to any point on Earth. This will result in practical wireless transmission of electrical power.

According to Grotz, although it was not until 1954-1959 when experimental measurements were made of the frequency that is propagated in the resonant cavity surrounding the Earth, recent analysis shows that it was Nikola Tesla who, in 1899, first noticed the existence of stationary waves in the Schumann Cav-

The power station for Tesla's Wardenclyffe Tower, c. 1916.

223

Tesla's experimental Colorado Springs tower in December 1899.

ity. Tesla's experimental measurements of the wave length and frequency involved closely match Schumann's theoretical calculations. Some of these observations were made in 1899 while Tesla was monitoring the electromagnetic radiations due to lightning discharges in a thunderstorm which passed over his Colorado Springs laboratory and then moved more than 200 miles eastward across the plains.

In his *Colorado Springs Notes*, Tesla noted that these stationary waves "...can be produced with an oscillator," and added in parenthesis, "This is of immense importance." The importance of his observations is due to the support they lend to the prime objective of the Colorado Springs laboratory. The intent of the experiments and the laboratory Tesla had constructed was to prove that wireless transmission of electrical power was possible.

According to the International Tesla Society, Schumann Resonance is nalogous to pushing a pendulum. A working Tesla wireless power system could create pulses or electrical disturbances that would travel in all directions around the Earth in the thin membrane of non-conductive air between the ground and the ionosphere. The pulses or waves would follow the surface of the Earth in all directions expanding outward to the maximum circumference of the Earth and contracting inward until meeting at a point opposite to that of the transmitter. This point is called the anti-pode. The traveling waves would be reflected back from the anti-pode to the transmitter to be reinforced and sent out again.

At the time of his measurements Tesla was experimenting with and researching methods for "...power transmission and transmission of intelligible messages to any point on the globe." Although Tesla was not able to commercially market a system to transmit power around the globe, modern scientific theory and mathematical calulations support his contention that the wireless propagation of electrical power is possible and a feasible alternative to the extensive and costly grid of electrical transmission lines used today for electrical power distribution.

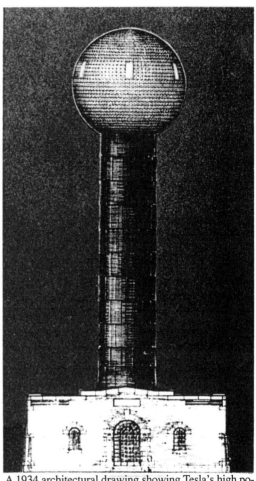

This system would essentially require a network of power towers to pulse energies into the atmosphere. These towers could be crystalline granite obelisks, exactly as we find in ancient Egypt and ancient Ethiopia.

In the next chapter we will discuss the alleged power system of Atlantis, the power system of the Gods—a power system that was similar to Tesla's, but used monolithic granite towers…

A 1934 architectural drawing showing Tesla's high potential terminal and powerhouse. This illustration was included in Tesla's beam weapon proposal to the U.S. government.

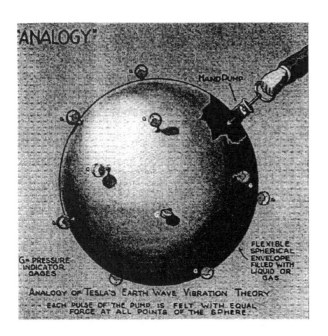

Power through the Earth as described by Tesla in the February 1919 issue of Electrical Experimenter. First it describes an "Analogy" to Tesla's proposed world-wide system utilizing towers and wireless electricity and then the "Realization" of such a system using Tesla's inventions. This system, we propose, is virtually identical to the the "Power System of the Gods."

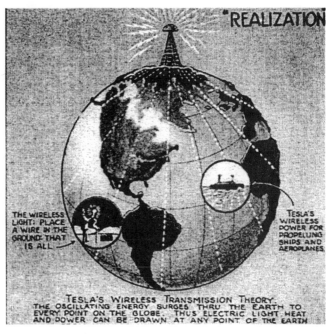

226

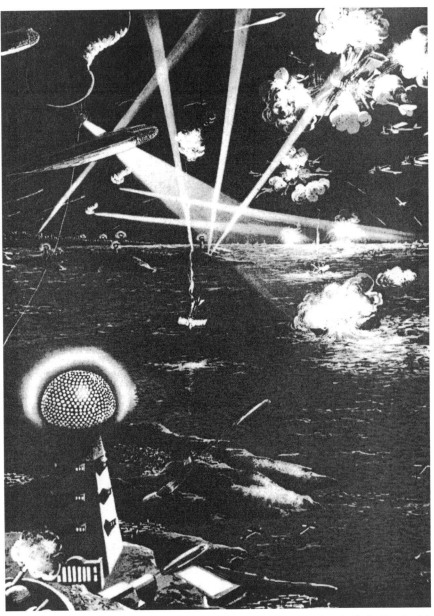

Tesla's wireless transmission tower in action, sending power to electrical airships in this illustration for the Feb. 1922 issue of *Science and Invention*. This is essentially the same system as we are proposing is the "Power System of the Gods."

The Wardenclyffe Tower being dynamited on July 4, 1917, in order to raise the $20,000 for Tesla's Waldorf-Astoria Hotel bill. With the destruction of this tower, Tesla's effort to create his wireless power system became only a dream.

Part 2
Chapter 3

The Terrible Crystals
of Atlantis

Arthur C. Clarke's Law of Revolutionary Ideas:

Every revolutionary idea—in science, politics, art, or whatever—
evokes three stages of reaction in a hearer:
1) It is completely impossible—don't waste my time.
2) It is possible, but it is not worth doing.
3) I said it was a good idea all along.

Words and language ...do not seem to play
any part in my thought processes.
—*Albert Einstein*

The Terrible Crystals of Atlantis
   The subject of Atlantis and the Power System of the Gods starts with the knowledge of Tesla's wireless transmission of power and then moves on to more subtle information, that of "psychic" sources. We previously drew upon the strange "experiences" of the astral-traveling Colonel Fawcett in Brazil. Now we examine more detailed sources on these "power towers of Atlantis."
   We will take them in chronicalogical order, from the older material to the more recent. We will examine three main sources on the subject: the strange "Phylos the

229

The "Unfed Light" in an Atlantean temple, according to "Phylos" in *An Earth Dweller's Return* (1940).

Thibetan" books; various Edgar Cayce readings; and the material from the Unarius group of the San Diego, California area.

This book, for the sake of expediency, will dispense with arguments for or against ancient aircraft or the existence of a former continent in the Atlantic Ocean called Atlantis. Rather, for the sake of the discussion at hand, we will postulate that an ancient technology that included electricity, hard metals, sonics and even aircraft, was used approximately six to sixteen thousand years ago. We can also assume that such a nation existed, capable of having such high technology, and that it stretched from the Northeast Africa across the Middle East to ancient India. It also apparently existed in a now submerged area of the Atlantic on the legendary island of Atlantis, or Poseid, as well as in the Yucatan and other parts of Central and South America. Discussions of Egypt and Ethiopia will wait, but let us now turn to the esoteric topic of the fascinating "crystals of Atlantis."

The knowledge of these "terrible crystals" and their strange technology comes largely from metaphysical texts such as Edgar Cayce, the Theosophical Society, the Lemurian Fellowship and other similar groups. In this book we will focus largely on two sources, that of Edgar Cayce and the books *A Dweller On Two Planets* and *An Earth Dweller Returns*.

The Incredible Maxt Tower of Atlantis

The power-towers of Atlantis, and the airships that drew power from them, were featured in great detail in two unusual books, *A Dweller On Two Planets*[71] and *An Earth Dweller Returns*.[72]

A Dweller On Two Planets was supposedly first dictated in 1884 by "Phylos the Thibetan" to a young Californian named Frederick Spencer Oliver who wrote the dictations down in manuscript form in 1886. The book was not published until 1899, when it was finally released as a book. In 1940, the sequel, *An Earth Dweller Returns*, was published by The Lemurian Fellowship of Ramona, California. Also accredited to "Phylos the Thibetan" this book was allegedly dictated to Beth Nimrai. Both books are the long and complicated history of a number of persons and the karma created by each of them during their many lives, especially the karmic relationships and events of the "amanuensis" Frederick Spencer Oliver and his different lives as Rexdahl, Aisa and Mainin with the many lives of "Phylos" as Ouardl, Zo Lahm, Zailm and Walter Pierson.

Both books are a complicated and often difficult to follow collection of past lives and the cycles of karma and rebirth between no less then eight people, men and women, and including Beth Nimrai, the amanuensis of the later book *An Earth Dweller Returns*. This book is largely an attempt to correct and clarify much of the material in *A Dweller On Two Planets* and both books contain a great deal of detailed information on the life, times, culture and technology of ancient Atlantis, including the airships which were called Vailxi in plural and Vailx in singular.

231

The electric carriage system used in the city of Caiphul in Atlantis, according to "Phylos."

A Dweller On Two Planets has remained a popular occult book for nearly a century largely because it contains detailed descriptions of devices and technology which were unquestionably well in advance of the time frame in which it was written. As the cover of one of the editions of the book states "One of the greatest wonders of our times is the uncanny way in which *A Dweller On Two Planets* predicted inventions which modern technology fulfilled after the writing of the book."[71]

Among the inventions and devices mentioned in both books are Air Conditioners, to overcome deadly and noxious vapors; Airless Cylinder Lamps, tubes of crystal illuminated by the "night side forces"; Electric Rifles, guns employing electricity as a propulsive force (rail-guns are a similar, and very new invention); Mono-Rail Transportation; Water Generators, an instrument for condensing water from the atmosphere; and the Vailx, an aerial ship governed by forces of levitation and repulsion.

Much of the wording and terms are identical to those in the Edgar Cayce readings, such as "night side forces" and the term "Poseid" for Atlantis. While verification of any of the information in both books is impossible, the material is fascinating and of definite interest to any student of the vimanas of ancient India. In chapter two of *A Dweller On Two Planets* the hero, Zailm (an earlier incarnation of Phylos and Walter Pierson), visits Caiphul, the capital of Atlantis, and views many wonderful electronic devices and the monorail system.

In chapter four the electromagnetic airships of Atlantis are introduced along with radio and television (don't forget, this book was written in 1886). It is explained that the airships, similar to zeppelins, but more like a cigar-shaped airship, are electro-magnetic-gravitational and are capable of entering the water as a submarine or traveling through the air. Later, in chapter sixteen, Zailm takes a journey via Vailx to "Suern" which is apparently ancient India or thereabouts.

In chapter eighteen Zailm visits "Umaurean" (present day American) colonies of Poseid. In a fascinating portion of the book, the Vailx stops for the night to visit a building on the summit of the Tetons. According to the text, "On the tallest of these had stood, perhaps for five centuries, a building made of heavy slabs of granite. It had originally been erected for the double purpose of worship of Incal (the Sun, or God), and astronomical calculations, but was used in my day as a monastery. There was no path up the peak, and the sole means of access was by vailx."

Frederick Spencer Oliver then alleges in a break in the story that such massive, granite slab-walls were discovered in 1886 by a Professor Hayden, allegedly the first person to climb Grand Teton. Whether such massive granite slabs, certainly in poor condition and probably thought to be naturally occurring, do indeed exist on or near the summit of Grand Teton, I have no way of knowing.

Afterwards they visit the ancient copper mines of the Lake Superior region (which do indeed exist and are archaeological fact, though not satisfactorily explained) and then return to Poseid, making part of the journey underwater.

Back in Atlantis (Poseid) Zailm makes the mistake of getting involved with two women at the same time and karmic repercussions are severe when, about to marry one of the women, the other exposes him and tragedy follows when both women are killed. One commits suicide and the other stands in the Maxin Light, a kind of super energy beam in the center of the great temple, analogous to that mentioned in the similar Edgar Cayce reading (440-5; Dec. 20, 1933).

This Maxin Light apparently had something to do with the giant energy towers and "terrible crystals" of Atlantis. Perhaps it was an giant arc of power coming off of one of the towers. These towers would also have func-

Dr. Robert D. Stelle of the Lemurian Fellowship.

233

The incredible Maxt Tower of Atlantis, according to "Phylos" in *An Earth Dweller's Return* (1940).

tioned in a manner similar to that designed by the great inventor Tesla.

In chapter eighteen, Zailm speeds away in his private Vailx and wanders for a time searching for gold in South America, using an electronic mineral detector, a water generator and an electric rifle. While searching for gold he is trapped in a small cavern by the evil priest Mainin (who is an early incarnation of Frederick Spencer Oliver, the amanuensis of the book) and dies.

A few incarnations later, Zailm (Phylos) is taken astral traveling to Venus, hence the title of the book, *A Dweller On Two Planets*. This part of the book is somewhat reminiscent of the Hari Krishna publication *Easy Journey To Other Planets* by Swami Prabhupada.[77]

In the second book, *An Earth Dweller's Return*,[72] much of the text is used to explain elements of *A Dweller On Two Planets* that had been left unexplained, particularly the karmic relationship between Phylos himself and Frederick Spencer Oliver. However a great deal of this book goes into the science of Atlantis including the cause of gravitational attraction; heat, magnetism and motion, transmutation of matter, the Maxin Light, another energy tower known as the Maxt, airless cylinder light, levitation and much more.

Part Five of the book is entitled "Description Of A Journey By Vailx." According to the text, "the Atlantean vailx was an air vessel motivated by currents derived from the Night Side of Nature.

"Altitude was dependent wholly upon pleasure. For this reason wide views were possible with a great variety of scenery. The rooms of the vailx were warmed by *Navaz* (Night Side of Nature) forces and furnished with the proper density of air by the same means. So rapidly did the aspect of things change beneath, that the spectator, looking backwards, gazed upon a dissolving view.

"The currents, derived from the Night Sides of Nature, permitted the attainment of the same rate of speed as the diurnal rotation of the earth. For example, suppose we were at an altitude of ten miles and that the time was the instant of the sun's meridian. At that meridian moment, we could remain indefinitely bows on, while the earth revolved beneath at approximately seventeen miles per minute. Or the reverse direction keys could be set, and our vailx would rush away from its position at the same almost frightful speed— frightful to one unused to it, but not so to the returning Atlanteans who, in the Aquarian Age to come, will travel the highways of the land, sea and air without a thought of fear."[72]

During the trip, the Vailx is beset by a storm: "The repulse keys were set, and presently we were so high in the air that all about our now closed ship were cirrus clouds—clouds of hail held aloft by the uprising of the winds which were severe enough to have been dangerous had our vessel been propelled by wings, fans, or gas reservoirs.

"But as we derived our forces of propulsion and repulsion from Nature's Night Side, or in Poseid phraseology, from *Navaz*, our long white aerial spindles feared no storm however severe. . . . The evening had not far advanced when it

An Atlantean Vailx in a storm (note "sparks"), from *A Dweller' On Two Planets* (1886).

was suggested that the storm would most likely be wilder near the earth, and so the repulse keys were set to a fixed degree, making nearer approach to the ground impossible as an accidental occurrence."

The chapter continues to speak of the journey, mentioning the destination, Suernis (India) and the air dispensers with wheels and pistons that pressurize the cabin. In Section 418 of the text it states, "The vailx used was about the middle traffic size. These vessels were made in four standard lengths; number one, about twenty-five feet; number two, eighty feet; number three about one hundred fifty-five feet; while the largest was approximately three hundred feet in length.

"These long spindles were round, hollow needles of aluminum, comprising an outer and an inner shell between which were placed many thousands of double 'T' braces, an arrangement productive of intense rigidity and strength. Other partitions made other braces of additional resistant force. From amidships the vessels tapered toward either end to sharp points. Most vailxi were provided with an arrangement which allowed an open promenade deck at one end. The vailx which Zailm used was about fifteen feet and seven inches in diameter.

"Crystal windows of enormous resistant strength were arranged in rows like port holes along the sides, with a few on top and several others set in the floor, thus affording a view in all directions."[72]

What is fascinating in reading descriptions of so-called Atlantean Vailxi in these books, as well as their brief descriptions in the Edgar Cayce readings, is their similarity to the descriptions of vimanas in ancient Indian texts and to a

236

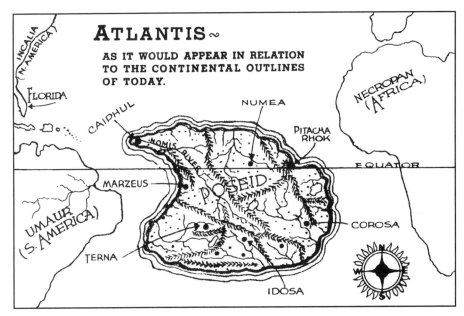

ATLANTIS ∾

AS IT WOULD APPEAR IN RELATION
TO THE CONTINENTAL OUTLINES
OF TODAY.

A map of Atlantis, according to "Phylos" in *An Earth Dweller's Return* (1940).

number of UFO craft seen in present times, including the late 1800s.

As the back cover paragraph on *A Dweller On Two Planets* points out, this book, and its description of a long, cylindrical, cigar-shaped aircraft, is a haunting premonition of not only many UFOs seen today, but of a type of craft that may yet be produced by a manufacturer in the near future!

At least several types of vimana aircraft are described by the ancient texts as circular in characteristics, which can either mean it is discoid, as in the mercury vortex type of propulsion, or it is cylindrical, as in the Vailx type of craft allegedly of Atlantis. That some of the UFO sightings of similar type craft of the past forty years might somehow be Atlantean or Indian-type Vailx or Vimana is a fantastic notion that apparently has never been considered by either the scientific community or by current UFO investigators.

The Great Crystal of Edgar Cayce

In a similar manner to the Phylos books is the "psychic" information from Edgar Cayce and the Association for Research and Enlightenment (A.R.E.) in Virginia Beach, Virginia, that was founded as a library and center to study Cayce's many dictations.

Known as the "sleeping clairvoyant," Edgar Cayce was born on March 18, 1877, on a farm near Hopkinsville, Kentucky. Even as a child he displayed powers of perception which seem to extend beyond the normal range. In 1898 at the age of twenty-one he became a salesman for a wholesale stationery

Edgar Cayce, c. 1930.

company and developed a gradual paralysi of the throat muscles which threatened the los of his voice. When doctors were unable to find a cause for the strange paralysis, he began to see a hypnotist. During a trance, the first of many, Cayce recommended medication and manipulative therapy which successfully re stored his voice and cured his throat trouble

He began doing readings for people mostly of a medical nature, and on October 9 1910, *The New York Times* carried two pages of headlines and pictures on the Cayce phe nomenon. By the time Edgar Cayce died on January 3, 1945, in Virginia Beach, Virginia he left well over 14,000 documented steno graphic records of the telepathic-clairvoyan statements he had given for more than 8,000 different people over a period of 43 years. These typewritten documents are referred to as "readings." Impor tant to our discussion in this book is that many of these "readings" concern Atlantis, persons' former lives in Atlantis, and the airships and motive powei used in Atlantis.[65]

In reading 2437-1; Jan. 23, 1941, Cayce told his subject: "...[I]n Atlantean land during those periods of greater expansion as to ways, means and man ners of applying greater conveniences for the people of the land—things of transportation, the aeroplane as called today, but then as ships of the air, for they sailed not only in the air but in other elements also."

A number of persons who came to Cayce for individual life readings were according to Cayce's reading, once navigators or engineers on these aircraft "[I]n Atlantean land when there were the developments of those things as made for motivative forces as carried the peoples into the various portions of the land and to other lands. Entity a navigator of note then." (2124-3; Oct. 2 1931)

"...[I]n Atlantean land when peoples understood the law of universal forces entity able to carry messages through space to the other lands, guided crafts of that period." (2494-1; Feb. 26, 1930)

Cayce called the motive power used in these vessels the "nightside of life." "[I]n Atlantean land or Poseidia—entity ruled in pomp and power and in understanding of the mysteries of the application of that often termed the nightside of life, or in applying the universal forces as understood in that pe riod." (2897-1; Dec. 15, 1929)

"...[I]n Atlantean period of those peoples that gained much in understand ing of mechanical laws and application of nightside of life for destruction." (2896-1; May 2, 1930)

238

Cayce speaks of the use of crystals or "firestones" for energy and related applications. He also speaks of the misuse of power and warnings of destruction to come:

"[I]n Atlantean land during the periods of exodus due to foretelling or fore-ordination of activities which were bringing about destructive forces. Among those who were not only in Yucatan but in the Pyrenees and Egyptian land, for the manners of transportation and communications through airships of that period were such as Ezekiel described at a much later date." (4353-4; Nov. 26, 1939. See *Ezekiel* 1:15-25, 10:9-17 RSV.)

"...[I]n Atlantis when there were activities that brought about the second upheaval in the land. Entity was what would be in the present the electrical engineer—applied those forces or influences for airplanes, ships, and what you would today call radio for constructive or destructive purposes." (1574-1; April 19, 1938)

"...[I]n Atlantean land before the second destruction when there was the dividing of islands, when the temptations were begun in activities of Sons of Belial and children of the Law of One. Entity among those that interpreted the messages received through the crystals and the fires that were to be the eternal fires of nature. New developments in air and water travel are no surprise to this entity as these were beginning development at that period for escape." (3004-1; May 15, 1943)

"...[I]n Atlantean land at time of development of electrical forces that dealt with transportation of craft from place to place, photographing at a distance, overcoming gravity itself, preparation of the crystal, the terrible mighty crys-

An Atlantean Vailx enters the water, from *A Dweller' On Two Planets* (1886).

tal; much of this brought destruction." (519-1; Feb. 20, 1934)

"...[I]n city of Peos in Atlantis—among people who gained understanding of application of nightside of life or negative influences in the earth's spheres of those who gave much understanding to the manner of sound, voice and picture and such to peoples of that period." (2856-1; June 7, 1930)

"...[I]n Poseidia the entity dwelt among those that had charge of the storage of the motivative forces from the great crystals that so condensed the lights the forms of the activities, as to guide the ships in the sea and in the air and in conveniences of the body as television and recording voice." (813-1; Feb. 5 1935)

The use of crystals as an important part of the technology is mentioned in a long reading from Dec. 29, 1933:

"About the firestone—the entity's activities then made such applications as dealt both with the constructive as well as destructive forces in that period It would be well that there be given something of a description of this so that it may be understood better by the entity in the present.

"In the center of a building which would today be said to be lined with nonconductive stone—something akin to asbestos, with ...other nonconductors such as are now being manufactured in England under a name which is well known to many of those who deal in such things.

"The building above the stone was oval; or a dome wherein there could be ...a portion for rolling back, so that the activity of the stars—the concentration of energies that emanate from bodies that are on fire themselves, along with elements that are found and not found in the earth's atmosphere.

"The concentration through the prisms or glass (as would be called in the present) was in such manner that it acted upon the instruments which were connected with the various modes of travel through induction methods which made much the [same] character of control as would in the present day be termed remote control through radio vibrations or directions; though the kind of force impelled from the stone acted upon the motivation forces in the crafts themselves.

"The building was constructed so that when the dome was rolled back there might be little or no hindrance in the direct application of power to various crafts that were to be impelled through space—whether within the radius of vision or whether directed under water or under other elements, or through other elements.

"The preparation of this stone was solely in the hands of the initiates at the time; and the entity was among those who directed the influences of radiation which arose, in the form of rays that were invisible to the eye but acted upon the stones themselves as set in the motivating forces—whether the aircraft were lifted by the gases of the period; or whether for guiding the more-of-pleasure vehicles that might pass along close to the earth, or crafts on the water or under the water.

John Worrell Keely in his Philadelphia laboratory, circa 1890.

"These, then, were impelled by the concentration of rays from the stone which was centered in the middle of the power station, or powerhouse (as would be the term in the present).

"In the active forces of these, the entity brought destructive forces by setting up—in various portions of the land—the kind that was to act in producing powers for the various forms of the people's activities in the same cities, the towns, and the countries surrounding same. These, not intentionally, were tuned too high; and brought the second period of destructive forces to the people of the land—and broke up the land into those isles which later became the scene of further destructive forces in the land.

"Through the same form of fire the bodies of individuals were regenerated; by burning—through application of rays from the stone—the influences that brought destructive forces to an animal organism. Hence the body often rejuvenated itself; and it remained in that land until the eventual destruction; joining with the peoples who made for the breaking up of the land—or joining with Belial, at the final destruction of the land. In this, the entity lost. At first it was not the intention nor desire for destructive forces. Later it was for ascension of power itself.

"As for a description of the manner of construction of the stone: we find it was a large cylindrical glass (as would be termed today); cut with facets in such manner that the capstone on top of it made for centralizing the power or force that concentrated between the end of the cylinder and the capstone itself. As indicated, the records as to ways of constructing same are in three places in the earth, as it stands today: in the sunken portion of Atlantis, or Poseidia, where a portion of the temples may yet be discovered under the slime of ages of sea water—near what is known as Bimini, off the coast of Florida. And (secondly) in the temple records that were in Egypt, where the entity acted later in cooperation with others towards preserving the records that came from the land where these had been kept. Also (thirdly) in records that were carried to what is now Yucatan, in America, where these stones (which they know so little about) are now—during the last few months— being uncovered." (440-5; Dec. 20, 1933)[65]

What Cayce was apparently the Tesla-type system using an obelisk with a capstone to transmit power to the world—the terrible crystal. Another power source may have been the giant maser that Christopher Dunn theorizes was in operation at Giza.

Electric Resonance for Power Generation

The use of resonance in Tesla system is key to system working, he claimed. Another contemporary of Tesla was John Worrell Keely, an inventor from Philadelphia who claimed that resonance was the key to all sorts of things including power generation, anti-gravity and free energy. Keeley (1837-1898) was the inventor of such mysterious devices as the "Vibrodyne Motor," the "Compound Disintegrator," and the "Provisional Engine." Much of his work was kept a secret, though he made frequent demonstrations to eager investors. Not much happened in the end, and Keeley has been branded a fraud artist.

We may never know the validity of Keeley's inventions, but Tesla—the world's greatest inventor—believed that resonance was key to power generation and transmission.

The following article was printed in the Indian newspaper *The Hindu, Science & Technology Supplement*, November 20, 1997. It is a translation of a Russian article written by Konstantin Smirnov for the magazine *RIA Novosti*:

> When Dr. Andrei Melnichenko, a physicist specialising in electrodynamics in the city of Chekhov near Moscow, called our editorial office and described his invention, I did not believe him. But my mistrust did not perplex the inventor, and he offered to demonstrate his device.
>
> The device consists of several batteries and a small converter to change direct current into alternating current (220V, 50Hz) using an electric motor.
>
> The power of this motor is far greater than that of the power source. When a small plate with several assemblies is added to the chain of components and switched on, the motor begins to pick up speed in such a way that it would be possible to set an abrasive circle on it and sharpen

a knife.

In another experiment, a fan serves as the final component of the device. At first, its blades are slowly rotating but, after a special unit is connected in sequence with it, the fan immediately gains speed and makes a good 'breeze'. All this looked strange, primarily from the standpoint of the law of conservation of energy.

Seeing my perplexity, Melnichenko explained that the processes taking place in his device are simple enough, and are based on the phenomenon of electric resonance.

Despite the fact that this phenomenon has been known for more than a century, it is only rarely used in radio engineering and communications electronics where amplification of a signal by many times is needed.

Resonance is not used much in electrical engineering and power generation. By the end of the last century, the great scientist Nicola Tesla used to say that without resonance, electrical engineering was just a waste of energy.

No one attached any importance to this pronouncement at that time. Many of Tesla's works and experiments, for instance the transmission of electricity by one unearthed wire, have only recently been explained.

The scientist staged these experiments a century ago, but it has only been in our days that S. V. Avramenko has managed to reproduce them. This also holds true for the transmission of electric power by means of electromagnetic waves and resonance transformers.

"My first experiments with high-frequency resonance transformers produced results which, to say the least, do not always accord with the law of the conservation of energy, but there is a simple mathematical and physical explanation of this," Melnichenko says.

"I have designed several special devices and electric motors which contain many of these ideas and which may help them achieve full resonance in a chain when it consumes energy only in the form of the thermal losses in the winding of the motor and wires of the circuits while the motor rotates without any consumption of energy whatsoever.

"This was shown during the demonstration," the inventor goes on to say. "The power, supplied to the motors, was less than was necessary for their normal operation! I have called the new physical effect transgeneration of electric power. Electric resonance is the principle underlying the operation of the device."

This effect can be very widely used. For instance, electric resonance motors may be employed in electric cars. In this case the storage batteries' mass is minimal.

The capacity, developed by an electric motor, exceeds the supplied electric power by many times, which may be used for devising absolutely autonomous propulsion power units—a kind of superpower plant under the hood.

The battery-driven vehicles, equipped with such power plants, would not need frequent recharging because, just as in the case of an ordinary engine, it would only need storage batteries for an electric start. All the results have been confirmed by hundreds of experiments with resonances in electric motors (both ordinary and special).

In special motors, it is possible to achieve the quality of resonance in excess of 10 units. The technology of their manufacture is extremely simple while the investments are minimal. The results are superb!

Electromechanics is only the first step. The next are statical devices, which are resonance-based electric power generators. For instance, a device, supplied at the input with power equal to that of three 'Energizer' batteries can make a 100-watt incandescent lamp burn at the exit.

The frequency is about 1 MHz. Such a device has a rather simple circuit, and is based on resonance. Using it, it is possible to by far increase the power factor of energy networks, and to drastically cut the input (reactive) resistance of ordinary transformers and electric motors.

But creation of fundamentally new, environmentally clean electric power generators is the most important application of electric resonance.

A resonance-based energy transformer will become the main element of such devices. The employment of conductors with very low active resistance—cryoelectrics—for their windings will make it possible to increase power by hundreds and thousands of times, in proportion to resonance qualities of the device.

The Russian Academy of Sciences, in its review says that the principle underlying the operation of the devices does not rouse doubts in theory and in practice, and that the work of the resonance-based electric systems is not in conflict with the laws of electrophysics.

According to the above article, resonance is key to proper power generation, transmission and reception. If crystalline granite towers were used for Tesla-type system they would need to be "tuned" like giant tuning forks. The would need to be "tuned" to the resonance of the electric power. Obelisk made of solid granite could fulfill this function...

Tesla and the Power System of Atlantis

According to the Unarius Academy of San Diego, California, Nikola Tesl. was the reincarnation of an Atlantean engineer and inventor who was respon sible for the energy supply first used to provide power on a now destroyec island in the Atlantic. In Unarius theory, from the great central pyramid in Atlantis, power beams would be relayed from reflectors on mountaintops inte the different homes where these power beams would be converted into light heat or even to cool the house.

According to Unarius, a round glass globe or sphere about a foot in diam-
er was filled with certain rare gases that would fluoresce and give off a soft
hite light, just as does a modern fluorescent light. Heating or cooling was
so quite simple: Air being made up of molecules of gases, electrical energy
f a certain frequency could be radiated through the air and converted into
at through "hysteresis" in the electromagnetic fields of the atoms.

The same proposition in reverse makes the air become cold. Similarly, the
mosphere on the earth is always converting certain electromagnetic energy
to heat. Speaking from the point of absolute zero (-495 degrees Fahrenheit),
l air on the surface of the earth is comparatively warm, even at the poles.

Cooling or heating the air at any given point means merely to decrease or
crease the "electromagnetic hysteresis." As a definition for a Pabst hyster-

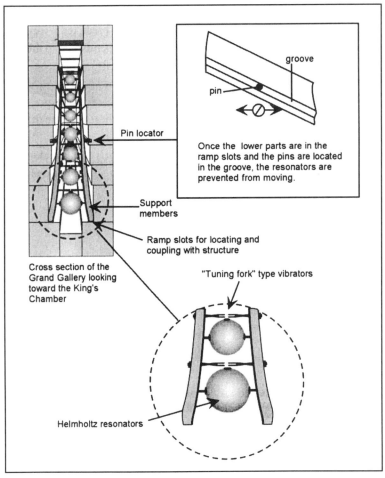

Christopher Dunn's diagram of the design and installation of the
Helmholtz resonators.

esis-synchronous motor, Unarius says that it is the "inductive principle of cos mic hysteresis," and adds that "The reference to 'hysteresis' is not the earth electronics definition, but rather an electromagnetic conversion process wherei cyclic (4th dim.) waveform-structures are transformed into lower (3rd dim waveform-structures."

Minoan homes are used as an example, where it is said that a small object foot or so square sitting on the floor of any room could be both the heater an the cooler. It would, according to the dictates of a thermostat, radiate certai energies into the room which would either slow down hysteresis and mak the air cooler or speed up hysteresis and make the air warmer; a far differer process than our present-day crude, clumsy, inefficient and enormous heat ing and cooling systems which must always either heat air in a furnace or coc it by means of refrigeration and, with a fan, blow it into the room through large duct.

The Atlantean Power System

Tesla's Atlantean power system, according to Unarius, was a huge rotatin squirrel-cage generator turned by a motor which was linked up to an elec tronic computer which was housed in a twenty-foot square metal box on th floor just above the generator. This computer automatically made and brok connections—with banks of power collector cells on the outside pyramid sur face in such a sequential manner that a tremendous oscillating voltage wa built up. On the ten-foot ball which stood atop the metal box, this oscillatin electricity discharged more than 600 feet straight up to a similar metal ba hanging down from the pyramid apex on a long metal rod.

Unarius compares the Atlantean-Tesla system to that of a 1900s scientis named Steinmetz, a friend of Tesla's. Steinmetz hurled thunder-bolts from two large metal spheres 100 feet apart in a manner which is somehow strangel similar to the process used in the Atlantean pyramid 16,000 years ago. Thi discharge across the two metal balls served as a tank-circuit, as it is called, anc again displays a similarity to our modern early-day wireless, a motor turnin a rimless rotary wheel from which protruded a number of spokes, actuall electrodes. As the wheel rotated about 2,000 rpm (rotations per minute), sizzling white spark jumped from the spokes to another electrode placed abou one-half inch away from the spokes. It was this spark-gap which created the necessary high-intensity voltage.

According to Unarius, on top of the Atlantean pyramid was a 50-foot meta column, something like a thick flagpole, which terminated in a circular banl of what looked like the spokes on a wheel. About 10 feet long and 16 inches in diameter, these spokes protruded at a number of irregular intervals, each on carefully sighted like a rifle, to a near or distant receiver. These spokes wer actually composed of an exotic mixture of metals and formed into a homoge neous, crystalline aggregate under extreme pressure and magnetic hysteresis

A scene from the 1961 movie *Atlantis, the Lost Continent.* A strange form of electricity, including crystals, is often a theme of the many Atlantis movies and novels.

ach rod or spoke then contained billions of tiny crystals; each one pointed, so ɔ speak, toward the outside flat of the rod. They absorbed energy and like a ɔy who'd eaten too much watermelon, they reached a certain capacity and ischarged their energy toward the outside end of the rod.

The net total of these charge and discharge oscillations was on the order of iillions of megacycles per second and as they functioned from the end of the ɔd, a beam of pure coherent energy emerged—and at the rate of more than 86,000 miles per second straight to a receiver, a beam of enormous power. Iow similar to our present first versions of the laser: A six-inch synthetic ruby ɔd, one inch in diameter and containing many chromium molecules; these hromium molecules were charged with electricity from an outside source of ɔndenser banks and other associated equipment which generated a high-requency impulse. As the chromium molecule atoms reached their satura-ion point, they discharged their energies which began to oscillate ping-pong ashion from each end of the optically-ground and silvered ends of the rod. Vhen this oscillating energy reached a certain point, it discharged through he more lightly silvered end in a single straight coherent beam of great inten-ity and power.

The power beams which emerged from the Atlantis pyramid were inter cepted by similar metallic rods of crystallized metal which, because they oscil lated in a similar manner and frequency, presented no resistance to the enor mous power of the beam. The beam then traveled straight through the rod o was broken up and separated into separate beams by a crystal prism, whicl again sent beams pulsating through crystalline rods and on a new tangent t another receiver.

In utilizing these power beams in a dwelling, a metal ball fitted on top of metal rod, like a small flagpole, contained a crystal of certain prismatic con figurations which directed the beam down through the hollow center of th rod to a disburser instrument which energized the entire house by means o induction so that the round milky-white crystal globes would glow with ligh motors turn, etc.

The Generator-Oscillator Banks

Unarius's technical description goes on to describe the generator-oscillato banks beneath the pyramid and the generation of the 'flame'. In the subterra nean chamber beneath the floor stood a motor-generator combination mounted on a vertical shaft. This piece of machinery "worked exactly similar to ou present day Pabst synchronous-hysteresis motor, that is, exactly in reverse t ordinary motors which have a rotor rotating inside fixed stationary field coil In the Pabst motor, the rotor is stationary and the metal field terminals rotat around it, similar to a squirrel cage.

"The Atlantean motor-generator combination works as follows: a hug externally-powered, (A.C.) alternating current motor rotated the squirrel cag which was actually a large number of extremely powerful high-gauss, high intensity magnets affixed to the metal frame which rotated around what woulc normally be the rotor which was made from a high-permeability, soft iror core. Wound around a large number of these poles were almost countles thousands of turns of insulated wire.

"These coils were, in turn, connected up to differen banks of cells on the outside skin surface of the pyra mid. The sequence of this wiring was such, that whe the magnets turned around the rotor, the cells and th magnetic currents so generated were in extremely rapi sequence which built up an extremely high-frequenc oscillating voltage which discharged across the tw balls which I described previously. The purpose of thi gap was to stabilize these oscillations under resistiv conditions in open air.

"Increasing the frequency increases the voltage o power which is why a laser beam can pierce a diamonc with less energy than would light a small flashligh

The Giza Power Plant.

The Giza Power Plant.

he energy from a five-foot long lightning bolt from a Tesla coil (500,000 Cycles ꞏer second) is less than two millionths of an ampere and would cause only a nild tingling sensation. A lightning bolt traveling from a cloud to the earth ꞏontains only enough energy to light a hundred-watt bulb for about thirty ꞏeconds."

Unarius contends that electronic scientists of today "are still a bit mixed up ꞏn the proposition of voltage versus frequency. They string 1/2 inch thick ꞏaminated cable across the countryside for hundreds of miles from tall steel ꞏowers and push electricity through these cables in far-away cities at voltages ꞏn excess of 300,000 and at only 60 cycles per second alternating frequency, ꞏvhereas a small pencil-thin power beam oscillating at hundreds of millions of ꞏimes per second could be reflected from tower to tower across country; one ꞏeam carrying sufficient power to energize the largest city."

ꞏrotective Metal Helmets

According to Unarius, and other esoteric groups that expound on ancient ꞏcience, in ancient Egypt, Mexico, and other lands where there were pyra-ꞏnids, the Egyptians and others tried to duplicate the round spoke-like wheel ꞏvhich glowed with a blue-white corona and which shot beams of intense light ꞏn different directions. The Egyptians topped their stone pyramid with a large ꞏall-like contrivance covered with small plates of pure polished gold in a scale-ꞏike manner; and as the earth turned, shafts of light were reflected in all direc-ꞏions.

Several thousands of years later, these metal balls with scales of gold had ꞏlisappeared, so had the alabaster white coating except for small sections near ꞏhe top, in order to use the smaller surface stones in nearby cities for building purposes.

The historic Egyptians wore in their temples and palaces a metallic head dress and woven metal scarves interwoven with threads of gold which hung down over their shoulders just as they did in ancient Atlantis when, after the scientists had gone, the Atlanteans started to worship the flame in the temple pyramid.

Unarius mentions that the metallic headdress plus a metallic robe was necessary to protect them from the strong electromagnetic field in the pyramid and that somehow the metallic headdress has arrived in our present modern time in the form of a scarf worn by women in a Catholic church, or the uraeus worn by the priest.

Here we see how the Egyptian gold headdress may have originated from the ancient Atlantean power station engineers, and it is fascinating to note that the celebrated Face On Mars is also wearing a similar protective helmet!

The Giant Giza Maser

The belief that the Great Pyramid at Giza was in fact a gigantic Maser sending power to a satellite in geosynchronous orbit has been championed by several authors, particularly by British engineer Christopher Dunn. Dunn is the author of the 1998 book *The Giza Power Plant: Technologies of Ancient Egypt*.[47] In this book, Dunn outlines his theories, and gives evidence for advanced machining and engineering knowledge in ancient Egypt. Another similar book is *The Giza Death Star* by Oklahoma physicist Joseph P. Farrell.[52]

Dunn claims that the Earth may be a giant power plant, and the many pyramids, obelisks, and megalithic standing stones may be part of this great "energy system." He says that the Great Pyra-mid was a giant power plant in the form of a maser and that harmonic resonators were housed in slots above the King's Chamber. He also theorized that there was a hydrogen explosion inside the King's Chamber that shut down the power plant's operation.

Says Dunn, "While modern research into architectural acoustics might predominantly focus upon minimizing the reverberation effects of sound in enclosed spaces, there is reason to believe that the ancient pyramid builders were attempting to achieve the opposite. The Grand Gallery, which is considered to be an architectural masterpiece, is an enclosed space in which resonators were installed in the slots along the ledge that runs the length of the Gallery. As the earth's viration flowed through the Great Pyramid, the resonators converted the energy to airborne sound. By design, the angles and surfaces of the walls and ceiling of the Grand Gallery, caused reflection of the sound and its focus into the King's Chamber. Although the King's Chamber was also responding to the energy flowing through the pyramid, much of the energy would flow past it. The design and utility of the Grand Gallery was to transfer the energy flowing through a large area of the pyramid into the resonant King's Chamber. This sound was then focused into the granite resonating cavity at sufficient amplitude to drive the granite ceiling beams to oscillation. These beams, in turn, compelled

Dunn's Egyptian satellite.

the beams above them to resonate in harmonic sympathy. Thus, the input of sound and the maximization of resonance, the entire granite complex, in effect, became a vibrating mass of energy.

"The acoustic qualities of the design of the upper chambers of the Great Pyramid have been referenced and confirmed by numerous visitors since the time of Napoleon, whose men discharged their pistols at the top of the Grand Gallery and noted that the explosion reverberated into the distance like rolling thunder."

Maintains Dunn, "While modern research into architectural acoustics might predominantly focus upon minimizing the reverberation effects of sound in enclosed spaces, there is reason to believe that the ancient pyramid builders were attempting to achieve the opposite. The Grand Gallery, which is considered to be an architectural masterpiece, is an enclosed space in which resonators were installed in the slots along the ledge that runs the length of the Gallery. As the earth's vibration flowed through the Great Pyramid, the resonators converted the energy to airborne sound. By design, the angles and surfaces of the walls and ceiling of the Grand Gallery, caused reflection of the sound and its focus into the King's Chamber. Although the King's Chamber was also responding to the energy flowing through the pyramid, much of the energy would flow past it. The design and utility of the Grand Gallery was to transfer the energy flowing through a large area of the pyramid into the resonant King's Chamber. This sound was then focused into the granite resonating cavity at sufficient amplitude to drive the granite ceiling beams to oscillation. These beams, in turn, compelled the beams above them to resonate in harmonic sympathy. Thus, the input of sound and the maximization of resonance, the entire granite complex, in effect, became a vibrating mass of energy.

"The acoustic qualities of the design of the upper chambers of the Gre Pyramid have been referenced and confirmed by numerous visitors since tl time of Napoleon, whose men discharged their pistols at the top of the Grar Gallery and noted that the explosion reverberated into the distance like ro ing thunder."

Dunn says that it is possible to confirm that the Grand Gallery indeed r flected the work of an acoustical engineer using only its dimensions, "Tl disappearance of the gallery resonators is easily explained, even though th structure was only accessible through a tortuously constricted shaft. The orig nal design of the resonators will always be open to question; however, there one device that performs in a manner that is necessary to respond sympathet cally with vibrations. There is no reason that similar devices cannot be create today. There are many individuals who possess the necessary skills to recr ate this equipment."[47]

According to Dunn, a Helmholtz resonator would respond to vibratior coming from within the earth, and actually maximize the transfer of energ The Helmholtz resonator is made of a round hollow sphere with a roun opening that is 1/10—1/5 the diameter of the sphere. The size of the sphe determines the frequency at which it will resonate. If the resonant frequenc of the resonator is in harmony with a vibrating source, such as a tuning fork, will draw energy from the fork and resonate at greater amplitude than tl fork will without its presence. It forces the fork to greater energy output tha what is normal. Unless the energy of the fork is replenished, the fork will lo: its energy quicker than it normally would without the Helmholtz resonato But as long as the source continues to vibrate, the resonator will continue t draw energy from it at a greater rate.

Dunn says that the Helmholtz resonator is normally made out of meta but can be made out of other materials. Holding these resonators in place inside the Gallery are members that are "keyed" into the structure by first being installed into the slots, and then held in the vertical position with "shot" pins that locate in the groove that runs the length of the Gallery.

Dunn now thinks that "The material for these members could have been wood, as trees are probably the most efficient responders to natural Earth

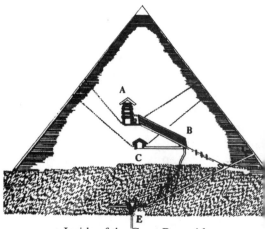

Inside of the Great Pyramid.

ounds. There are trees that, by virtue of their internal structure, such as cavi-ies, are known to emit sounds or hum. Modern concert halls are designed nd built to interact with the instruments performing within. They are huge nusical instruments in themselves. The Great Pyramid can be seen as a huge nusical instrument with each element designed to enhance the performance f the other. To choose natural materials, especially in the function of resonat-ng devices, would be a natural and logical decision to make. The qualities of vood cannot be synthesized."

What Dunn is describing seems fantastic—a giant beam of maser energy vas broadcast out of the pyramid to what calls his "Eye of Horus" satellite vhich then distributed the power to other locations on the earth—possibly to uned crystal towers.

Essentially, the preceding stories offer us a curious "what if?"

What if there was an ancient power system—a power system that both ;ave power to, and helped guide, the airships of the time?

What if there was a large-scale system that utilized crystal broadcast tow-rs to send power into the atmosphere?

As we will see in the next chapter, giant, pointed, crystal towers do in fact •xist. They can be seen in a number of cities around the world, expecially in :gypt. They are known to be ancient. The are called obelisks. How old they re, how they were quarried and transported, and even exactly what they vere for, is not really known for sure by modern scholars.

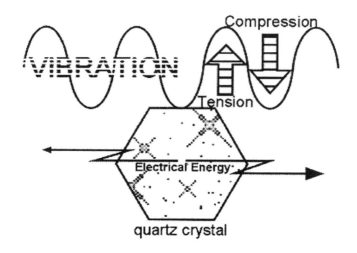

The piezoelectric effect.

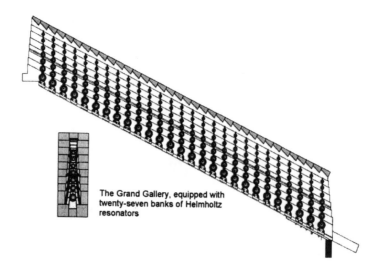

The Grand Gallery, equipped with
twenty-seven banks of Helmholtz
resonators

Christopher Dunn's diagram of the Grand Gallery Resonators.

Part 2
Chapter 4

The Crystal Towers of Egypt & Abyssinia

The priests told me that the Great Pyramid embodied all the wonders of Physics.
—Herodotus (350 BC)

Bassagordian's Basic Principle and Ultimate Axiom:
By definition, when you are investigating the unknown, you do not know what you will find or even when you have found it.

Weiler's Law:
Nothing is impossible for the person who doesn't have to do it himself.

The Mystery of Obelisks

The standard definition of an obelisk is a monolithic stone monument whose four sides, which generally carry inscriptions, gently taper into a pyramidion at the top. These massive, pointed shafts of polished granite were often capped with gold. Their size undoubtedly made them extremely difficult to move and raise.

The reason for obelisks, and their shape, has always been obscure. According to the 1997

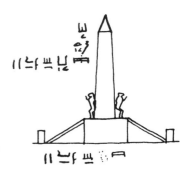

255

Grolier's Encyclopedia, "The ancient Egyptians usually erected them in pairs and associated them with the rays of the Sun, which increase in width as they reach the Earth. The earliest known examples, excavated at Abu Sir, Egypt, date from the Old Kingdom during the reign of Neuserre (2449-2417 BC)."

Possible phallus symbolism, as in church steeples, must have been speculated by at least a few Egyptologists. That obelisks may have symbolized the "pillar of the lord" or a shaft of sunlight have all been various theories. The fact is, no Egyptologist has come up with a really good explanation as to why tremendous effort was made to erect obelisks. We could almost state the current state of opinion as: "Hey, the Egyptians just liked to erect obelisks."

Obelisks weren't erected only in Egypt. A number of obelisks were erected during the "prehistory" of Ethiopia, others possibly elsewhere. More on those obelisks later. The real purpose of these obelisks, according to the thesis of this book, was as antennas for receiving and possibly for transmission. The dynastic Egyptians, however, used them for different purposes, mainly as monuments.

The Dynastic Egyptians who had lost this science, moved various obelisks from their original sites, currently unknown, and placed them as monuments in temples in Karnak, Heliopolis and a few other places.

However, without a doubt, the Dynastic Egyptians carved and erected some of the smaller obelisks. The larger obelisks were probably already fallen, or perhaps still in place, and then reused during dynastic times. Whether the ancient Egyptians had the technology to quarry, transport and erect huge obelisks is still a very serious debate today. While huge, erect obelisks are a simple archeological fact, how they came to be where they are is a major mystery.

Still unexplained are:
1. How were obelisks quarried?
2. How were obelisks removed from the quarry and transported?
3. How were obelisks moved onto the ships and later moved off?
4. How were obelisks transported to the erection site?

5. How were the obelisks erected?
6. What was the purpose of obelisks since it was such a tremendous effort to do all of the above?

While reasonable guesses and logical theories have been advanced for some of these questions, the exact techniques and various mysterious elements remain, as we shall see. Explanations such as stone balls, rollers, and hardened copper saws have their problems—read on!

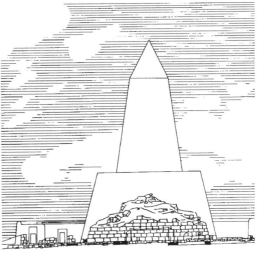
A drawing of a "Sun Temple" reconstructed.

The Mystery of Pyramidions and Sun Temples

In 1550 BC, more than 1,200 years after the Giza Pyramids were thought to have been erected (they may be older), ancient Egyptian pharaohs were building magnificent temples adorned with huge monuments that pointed to the sky.

An obelisk is a four-sided single piece of stone standing upright, gradually tapering as it rises and terminating in a small pyramid called a pyramidion. Obelisks were known to the ancient Egyptians as Tekhenu, a word whose

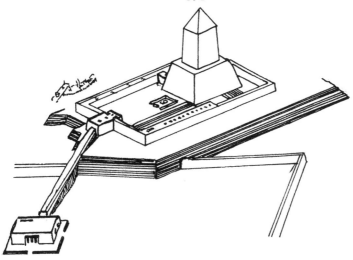
A drawing of a "Sun Temple" reconstructed.

257

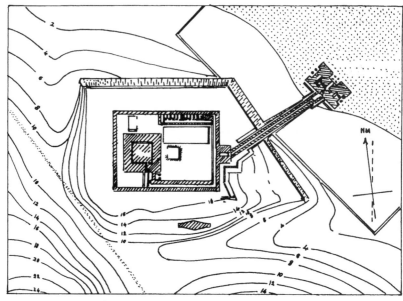

A drawing of a "Sun Temple" reconstructed.

derivation is unknown. When the Greeks became interested in Egypt, both
obelisks and pyramids attracted their attention. To the former they gave the
name "obeliskos," from which the modern name in almost all languages is
derived. Obeliskos is a Greek diminutive meaning "small spit"; it was applied
to obelisks because of their tall, narrow shape. In Arabic, the term is Messalah,
which means a large patching needle and again has reference to the object's
form.

Around the Heliopolis area, these monoliths were commonly of red granite from Syene and were dedicated to the sun god. In traditional dynastic Egypt
they were usually placed in pairs before the temples, one on either side of the
portal. Few actual temples with obelisks remain,
the main one being Luxor Temple.

Down each of the four faces, in most cases,
ran a line of deeply incised hieroglyphs and representations, setting forth the names and titles
of the pharaoh. The cap, or pyramidion, was
sometimes sheathed with copper or other metal.

According to traditional Egyptology, obelisks
of colossal size were first raised in the 12th dynasty. Of those still standing in Egypt, one remains at Heliopolis and two at Karnak, one from

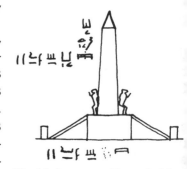

Thoth baboons worship an obelisk.

258

he time of Thutmose I and one of Queen Hatshepsut which is estimated to be)7.5 ft (29.7 m) high.

Queen Hatshepsut born c. 1482 BC), an Egyptian queen of the 18th dynasty, was the only woman to rule Egypt as a pharaoh. After the death (c.1504 BC) of her husband, Thutmose II, she assumed power, first as regent for his son Thutmosis III, and then (c. 1503 BC) as pharaoh. She is credited with erecting one of the obelisks at the Temple of Karnak in Luxor.

One of the theories of this book is that these, like all the famous obelisks, were erected many thousands of years earlier, by another culture. Being virtually indestructable, they later had temples built around them. Originally, we theorize, the obelisks

The obelisk at Karnak Temple, Luxor.

were polished granite without any hieroglyphs. During Queen Hatshepsut reign they were basically carved with hieroglyphs and dedicated.

According to American University of Cairo professor Labib Habachi, obelisks are the most often seen—and best known—of all the objects and structures of the ancient civilizations of the past, including all things Mesopotamian, Greek, Roman, or Egyptian.

Says Professor Habachi (who died in 1984), "Some of the smaller obelisks and fragments of larger ones are familiar to the numerous visitors of museums in various countries; larger ones which are still on their original sites are admired by the thousands of people who visit Egypt each year. Still others are seen by the crowds who pass through London, Paris, New York, Istanbul, and especially Rome, where there are more obelisks than in any other place."[45]

Obelisks were considered by the ancient Egyptians to be sacred to the sun god, whose main center of worship was at Heliopolis, the ruins of which lie in the district of Matariya near Cairo. Although the well-known obelisks date from the 20th century BC, such monuments seem to have been erected there in honor of the sun god in much earlier periods.

According to Habachi, a type of stone resembling the pyramidion of an obelisk was apparently considered sacred to the sun god even before the appearance of the first pharaoh in the First Dynasty (c. 3100-2890 BC). Such stones, known as ben or benben, were believed to have existed in Heliopolis from

The obelisk at Luxor Temple, Luxor.

time immemorial and were the fetish of the primeval god Atum (the setting sun) and the god Re or Re-Harakhti (the rising sun). Habachi is essentially saying that the use of obelisks is "predynastic." This book theorizes that all of the large obelisks are predynastic. There is no recognized scientific technique for dating when a granite rock was quarried as yet.

Benben stones were also associated with the Benu-bird, or phoenix. This creature, which begot itself, was thought to have come from the east to live in Heliopolis for 500 years and then to return to the east to be buried by the young phoenix which would in turn replace it in Heliopolis. According to one version of the tale, instead of being replaced the bird revived itself, and thus it was connected with the god of the dead. In some tombs, an image of the phoenix is shown among the gods.

In the pyramids of the last king of the Fifth Dynasty and the kings of the Sixth Dynasty (c. 2345-2181 BC), the walls of the burial chamber were decorated with Pyramid Texts, religious texts concerned with the welfare of the deceased. One text reads: "O Atum, the Creator. You became high on the height, you rose up as the benben-stone in the mansion of the 'Phoenix' in Heliopolis."

Habachi says that Pliny the Elder (A.D. 23-79), the Roman encyclopedist, wrote that obelisks were meant to resemble the rays of the sun. This comparison finds support in an inscription addressed to the sun god: "Ubenek em

The obelisk at Luxor Temple, Luxor.

Benben" ["You shine in the benben-stone."]. During the prosperous days of the 18th Dynasty (1570-1320 BC), and perhaps at other times, the pyramidions of obelisks were covered with gold or some other metal. The date at which obelisks were first erected is not known, but the kings of the Fifth Dynasty (2494-2345 BC), who were fervent worshipers of the sun god, may have been the earliest rulers to decorate the facades of their temples with pairs of such monuments.

Says Habachi:

Heliopolis, the city of the sun, was called by the ancient Egyptians Iunu, a name meaning "the pillar," and sometimes Iunu Meht, "the northern pillar." The name Iunu appears in the Bible as On; Heliopolis is the Greek name by which the city is generally known. Heliopolis was sacred to the sun god Re and his ennead, a group of nine associated gods. Other gods worshiped there included Kheperi, the scarab, and Shu, the god of the air. Obelisks were first erected at Heliopolis and the practice was continued throughout the pharaonic period. The majority of these obelisks have been removed or destroyed; the only one still standing there is that of Sesostris I (1971-1928 BC).

Ancient Thebes (modem Luxor) was known as Uast, "the scepter," or sometimes as Iunu Shemayit, "the southern pillar," or as "the Heliopolis of the south." Its main god, Amun, was represented in human form with a crown of tall feathers. He was later assimilated with

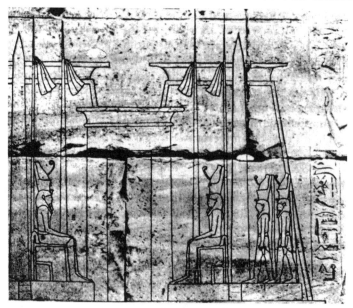

A ceremony featuring an obelisk.

262

Diagrams showing theoretical way to raise an obelisk.

Re and was known as Amun-Re, "King of the Gods." Because of this identification, obelisks were raised on his behalf. In Thebes, the center of his cult, numerous obelisks, including many of the largest, were erected in honor of Amun Re, at the time when the city was the capital of Egypt. Of its obelisks, only three survive; some were destroyed and a few were taken abroad.

Piramesse—that is, Per-Ramessu, "the domain of Rameses" became the capital of Egypt in the reign of Ramesses II (1304-1237 BC) and remained so under the succeeding Ramesside kings of the 19th and 20th dynasties (1320-1200 and 1200-1085 BC). It was embellished with a score of obelisks, for the most part fashioned by Ramesses II, although several made by earlier kings were taken over by him. Most of these obelisks were smaller than those of Thebes. The cults of the great gods, Re, Amun-Re, and Ptah of Memphis, were introduced in the new capital, and the names of these and other gods appear upon the obelisks Ramesses II erected there.

Elsewhere, according to Habachi, only rather small obelisks have been found. The inscriptions on these make it clear that they were erected in honor of local divinities who were either solar gods or associated with the solar cult. Two pairs of obelisks which were recovered from the ruins on the island of Elephantine near Aswan were dedicated to Khnum, the ram-headed god who fashioned mankind upon a potter's wheel. He was later associated with the sun god Re and was known as Khnum-Re. His obelisks were set up at the "altar of Re"—probably in a solar chapel. During the nineteenth century a pair of Ptolemaic obelisks dedicated to Isis was unearthed on the island of Philae. On them are mentioned the solar gods Atum and Amun Re. At Abu Simbel, a chapel of Re Harakhti adjacent to the Great Temple of Ramesses II, contained a pair of obelisks and other cult objects related to the sun god.

According to Habachi, an obelisk found at Minshah in Middle Egypt undoubtedly once stood at the neighboring religious center of Abydos. On this obelisk the king is called

Top: A scarab featuring the worship of an obelisk. Bottom: A vignette from *The Book of the Dead* featuring two obelisks used in a ceremony.

"beloved of Osiris," the god of the dead, who was the principal god of Abydos, although other deities also had cult places there. He says that there were two obelisks in Ashmunein, also in Middle Egypt, both dedicated to Thoth, god of writing and wisdom, and titulary deity of the place.

Among Thoth's many attributes was that of "representative of Re." The ibis and the baboon were sacred to Thoth, and the latter animals were often shown adoring the sun god. In the quarries of Gebel el-Ahmar near Cairo an obelisk is depicted standing between two baboons with their front legs raised in worship. The curious depiction of the Thoth baboons worshipping an obelisk is strange indeed. Did some sort of power or energy come out of the obelisk?

Says Habachi:

> An inscription on a fragment of an obelisk from Horbeit in the eastern part of the Delta mentions Osiris and his sacred bun, the Mnevis, known as "the living soul of Re." At Athribis in the center of the Delta the pedestals and a few fragments of the shafts of two obelisks still remain. On one fragment, the king is shown with local divinities, one of which is Atum. A number of obelisks were also raised in honor of Atum, a local god of Sais, which was a political center in the Delta and capital of the country during the Saite Period (664-525 BC).
>
> The kings who erected obelisks were usually described on them as beloved of various local and solar gods, and in many cases the king was shown in close relationship to these divinities. One text from an obelisk describes the king as "appearing like Harakhti, beautiful as King of the Two Lands like Atum," and a second, as "the one whom Atum made to be King of the Two Lands and to whom [he] gave Egypt, the desert, and foreign lands."

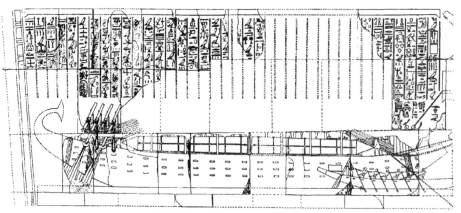

Moving one of Queen Hatsepsut's obelisks.

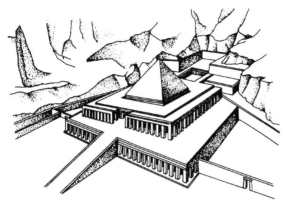

A drawing of a "Sun Temple" reconstructed.

Cleopatra's needle

On some obelisks there are references to royal victories, but these are rarely actual historical events. The boasts on most of the obelisks erected by Ramesses II are particularly suspect. On one, this king is commemorated as "the one who defeats the land of Asia, who vanquishes the Nine Bows, who makes the foreign lands as if they were not." On another it is said, "His power is like that of Monthu [the god of war], the bull who tramples the foreign lands and kills the rebels." The king is described as recipient of tribute, noble governor, brave, and vigilant.

If the claims of Ramesses II are not justified, those of his predecessor Tuthmosis III carry greater weight. In celebration of the great victory which Tuthmosis III gained over his powerful enemies in Asia, he erected two obelisks at the Temple of Karnak; the upper part of one of these survives in Istanbul. On one side of it, the king is spoken of as "the lord of victory, who subdues every [land] and who establishes his frontier at the beginning of the earth [the extreme south] and at the marshland up to Naharina [in the north]." On another side, he is said to have crossed the Euphrates with his army to make great slaughter. This crossing of the river was a great achievement, equaled only by his grandfather Tuthmosis I. It provided sufficient justification for the erection of the obelisks.

Yet another reason for setting up obelisks is indicated on the Istanbul obelisk, where Tuthmosis III is described as "a king who conquers all the lands, long of life and lord of Jubilees." Beginning in the thirtieth year of a king's reign, and every three years thereafter, a festival of renewal was

265

celebrated. On the occasion of these jubilees, the kings set up obelisks. The obelisk of Queen Hatshepsut (1503-1482 BC), which still stands at the Temple of Karnak, describes her as "the one for whom her father Amun established the name 'Makare' upon the Ashed-tree [a tree of eternity] in reward for this hard, beautiful, and excellent monument which she made for her First jubilee." However, since Hatshepsut reigned only about twenty years, she evidently celebrated her jubilee much earlier.

Habachi says that obelisks were always regarded by the ancient Egyptians as a symbol of the sun god related to the *Benben*, but during certain periods they were looked upon as being themselves occupied by a god and thus entitled to offerings. This is the case with four obelisks erected by Tuthmosis III in the Temple of Amun-Re at Karnak. In an inscription there, the king recorded the establishment of new feasts and offerings which he instituted for the four obelisks, dedicating 25 loaves of bread and jars of beer to each of them daily. When an obelisk was erected, scarabs showing the king kneeling in adoration before it were issued. A vignette accompanying the 15th chapter of the *Book of the Dead*, a guide to the dead in their travels through the Underworld, is entitled "Adoring Re-Harakhti when he rises in the eastern horizon of the sky." In the scene are two priests, one reciting from a roll of papyrus which he holds, the other making offerings to two obelisks which embody Re-Harakhti.

In addition to large obelisks, smaller obelisks—or rather obeliskoid objects—

were sometimes placed in front of tombs. These objects were inscribed on only one face with the name and the main title of the tomb owner. Some bear prayers addressed to the gods of the dead on behalf of the tomb owner. During the Old Kingdom, the kings erected pyramids to serve as their burial places. In fact, the pyramids were but a focus of a large funerary complex with a mortuary temple at the base of the pyramid connected by means of a causeway to a valley temple on the edge of the cultivated area. The kings of the Fifth Dynasty, already mentioned as especially devoted to the sun god, added to their pyramid complexes solar temples in which a gigantic obelisk was the main feature.

These strange "solar temples" with their obelisks and causeways look a lot like a ceremonial imitation of an actual power station. What sort of rituals were the dynastic Egyptians using to worship these ancient towers? Why were they thought to have some sort of divine power? How were they even quarried and erected? And why would anyone go through such tremendous effort to raise a seemingly non-functional monument?

The unfinished obelisk at the Aswan quarry.

267

Above: Drawings of obelisks done by Napoleon's artists.

268

How To Quarry An Obelisk

Today, quarrymen cut and carve granite using saws with diamond-edged blades and hardened steel chisels. Ultrasonic drilling is also used.

Modern Egyptologists assume that the ancient Egyptian quarrymen and stonemasons didn't have these modern tools. How, then, did they quarry and cut such clean lines in their obelisks and other monumental statuary? The quarrying, transporting, and lifting of gigantic blocks of stone has been an enduring mystery, one that has brought forth many fascinating theories, from the very simple to the very complex.

Even the lifting of obelisks has posed a great mystery and mainstream archeologists have had to admit that they don't really know how the obelisks were put into place, nor exactly how they were quarried either.

In March of 1999, the NOVA television documentary team went to Egypt to produce a special on obelisks entitled *Pharaoh's Obelisk.* The documentary team brought along an ancient-tools expert to find out how ancient Egyptians quarried huge pieces of granite for their obelisks. NOVA traveled to an ancient quarry in Aswan, located 500 miles south of Cairo. This is where the ancient Egyptians found many of the huge granite stones they used for their monuments and statues.

One of the most famous stones left behind is the Unfinished Obelisk, more than twice the size of any known obelisk ever raised. Quarrymen apparently abandoned the obelisk when fractures appeared in its sides. However, the stone, still attached to bedrock, gives important clues to how the ancients may have quarried granite.

In the NOVA documentary, archeologist Mark Lehner, a key member of the expedition, crouches in a granite trench that abuts one side of the Unfinished Obelisk. Lehner holds a piece of dolerite similar to the kind that he and others believe Egyptian quarrymen used to pound out the trench around the edges of the obelisk. They then lifted the pulverized granite dust out of the

The obelisk at Luxor Temple.

trenches with baskets.

It is theorized that workers pounded underneath the obelisk until the monu ment rested on a thin spine. Lehner says that huge levers were probably used to snap the obelisk from its spine, freeing it so it could be carved more finely and transported.

Archeologists know that the ancient Egyptians had the skills to forge bronze and copper tools. In the documentary, stonemason Roger Hopkins takes up a cop per chisel, which works well when carving sandstone and limestone rock, to see if it might carve granite.

"We're losing a lot of metal and very little stone is falling off," observes Hopkins, which is hardly the desired result. Hopkins' simple experiment makes this much clear: The Egyptians needed better tools than soft bronze and copper chisels to carve granite.

NOVA then brought in Denys Stocks, who, as a young man, was obsessed with the Egyptians. For the past 20 years, this ancient-tools specialist has been recreating tools the Egyptians might have used. He believes Egyptians were able to cut and carve granite by adding a dash of one of Egypt's most common materials: sand.

"We're going to put sand inside the groove and we're going to put the saw on top of the sand," Stocks says. "Then we're going to let the sand do the cutting."

It does. The weight of the copper saw rubs the sand crystals, which are as hard as granite, against the stone. A groove soon appears in the granite. It's clear that this technique works well and could have been used by the ancient Egyptians.

But is this how the giant stone obelisks and other blocks were cut?

Hopkins' experience working with stone leads him to believe that one more ingredient, even more basic than sand, will improve the efficiency of the granite cutting: water. Water, Hopkins argues, will wash away dust that acts as a buffer to the sand, slowing the progress.

Adding water, though, makes it harder to pull the copper saw back and forth. While Hopkins is convinced water improves the speed of work, Stocks' measurements show that the rate of cutting is the same whether water is used or not.

Besides cutting clean surfaces on their granite, the Egyptians also drilled cylindrical holes into their stones. A hole eight inches in diameter was found drilled in a

NOVA tried but failed to raise an obelisk by the above method.

granite block at the Temple of Karnak.

"Even with modern tools—stone chisels and diamond wheels—we would have a tough time doing such fine work in granite," says Hopkins.

Stocks was brought along to test his theories about how the cores were drilled. Inspired by a bow drill seen in an ancient Egyptian wall painting, Stocks designs a homemade bow drill. He wraps rope around a copper pipe that the Egyptians could have forged. Hopkins and Lehner then pull back and forth on the bow, which is weighted from above. The pipe spins in place, rubbing the sand, which etches a circle into the

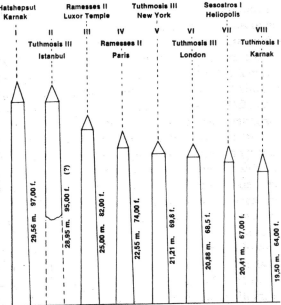

A comparison of known preserved obelisks.

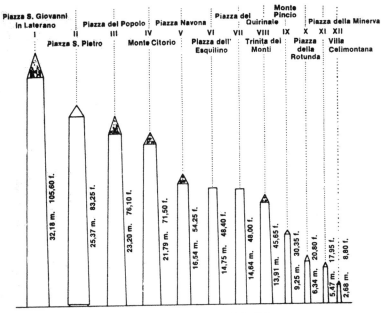

A comparison of known preserved obelisks.

The Washington Monument being hit by lighting.

stone. With the assistance of the sand, the turning copper pipe succeeds in cutting a hole into the granite slab.

With the aid of a bow drill and sand, the pipe has cut a circular hole into the stone. But how can the drillers get the central core out?

In the NOVA documentary, Stocks wedges two chisels into the circular groove. The core breaks off at its base. Stocks reaches in and plucks it out, leaving a hole behind not unlike the ones once cut by the Egyptians.

This may well be how Egyptian workers were able to cut through granite with copper saws with the addition of a little abrasive sand. Sure, it was slow, hard work, but it could be done. Stocks' experiments squelched the various speculations about the ancient Egyptians possessing superhard tools or being able to soften stone. However, C. Ginenthal writes that H. Garland and C.O. Bannister attempted to saw through granite back in the 1920s using essentially the same method employed by Stocks—but without success. Garland and Bannister wrote a book on their experiments, from which Ginenthal has provided the following quotation:

"A consideration of the [copper and abrasive cutting] process would seem to give support to the idea that a copper-emery [or other abrasive material] process might have been used by the first Egyptians, but the author [Garland] has proved by experiment the impossibility of cutting granite or diorite by any similar means to these. But the use of emery powder anointed with oil or turpentine, no measurable progress could be made in the stone whilst the edge of the copper blade wore away and was rendered useless, the bottom and sides of the groove being coated with particles of copper." (*Ancient Egyptian Metallurgy*, London, 1927, p. 95)

The Use and Reuse of Obelisks in Egyptian Times

Egypt as a civilization lasted for many thousands of years. Compared to our modern western civilization, only a few hundred years old, ancient Egypt has literally thousands of years of history. The beginnings of Egypt are said to go back to the "King Scorpion" (shades of the recent "Mummy" films) and the king known as Narmer. This time is generally thought to be around 3150 BC. Known as the Archaic Period, it is generally said to last up to the Early Dynastic Period starting in 3050 BC.

The Early Dynastic Period begins with the first "historical" pharoah, Menes, who is credited with "unifying" (at least symbolically) upper and lower Egypt.

272

He was followed by the pharoahs "Aha," "Djer," "Djet," and "Den." The Early Dynastic Period ended around 2575 BC and the Old Kingdom began. The pyramids are thought by mainstream Egyptologists to have been built at this time.

The Middle Kingdom begins around 2040 BC and the New Kingdom, which included Akhenaton, Nefertiti and Tutankamun, began in 1550 BC, ending in 1070 BC. The ancient Egyptian culture basically came to an end in 343 BC with its conquest by Persia. Alexander the Great conquered Egypt in 332 BC, beginning the Greco-Roman period.

But what concerns us here is obelisks and their antiquity. Did obelisks exist in the land of Egypt before "King Scorpion" in 3150 BC? The contention of this book is that they did indeed exist before that time, as did the Sphinx and the pyramids.

In fact, obelisks, along with various megalithic structures such as the pyramids, the Valley Temple of Chephren and the Osirion, may well be from a time many thousands of years before the currently accepted beginning of dynastic and predynastic Egypt. The theory presented here is that these structures are over 10,000 years old and have been occupied and reused by the dynastic Egyptians.

Obelisks Abroad

Many of these giant crystalline shafts have been carried from Egypt to other countries, notably one from Luxor, now in the Place de la Concorde, Paris, and Cleopatra's Needles in London and New York. Others are in Rome and Florence. In the United States two familiar structures of obelisk form (though not monoliths) are the Washington and the Bunker Hill monuments.

So popular were these monuments among the Roman emperors that 13 of them were taken to Rome. Cleopatra's Needles, named for the famous Egyptian queen (though having nothing to do with her, or any of the many other Cleopatras), are two ancient obelisks presented by the khedive of Egypt to Great Britain (1878) and the United States (1880). The British monument, 20.9 m (68.5 ft) high, is located on the Thames Embankment in London. The American one, 21.2 m (69.5 ft) high, stands in New York City's Central Park. The British installed their obelisk at its present location in 1878; the

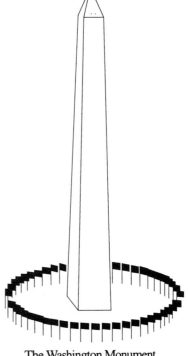

The Washington Monument.

Americans did so in 1881. Carved from rose-colored Syrene granite and inscribed with hieroglyphics, each weighs approximately 200 tons.

They had been previously erected c. 1500 BC in the city of Heliopolis by Thutmosis III. They may have been older than 1500 BC and erected for a second time at Heliopolis by Thutmosis III. They were again (?) moved by the Romans in 14 BC, Augustus ordered their removal to Alexandria to grace the grounds of the Caesareum.

The Washington Monument is an obelisk as well, though it is not a solid crystal obelisk as are most obelisks including Cleopatra's Needles.

George Washington personally selected the site of the nation's permanent capital in 1791, and the government was officially transferred there in 1800. Located close to the geographic center of the original 13 colonies, the area allotted measured 259 sq km (100 sq mi) and encompassed the existing port towns of Alexandria and Georgetown. The land west of the Potomac was returned to Virginia in 1846. Pierre Charles L'Enfant's design (1791) for the city, developed after 1801, was limited to the area south of the present Florida Avenue. It consisted of a physical framework for the siting of major government buildings (particularly the White House and Capitol) and a grid street pattern overlaid by broad radial avenues, with a series of squares and circles reserved for monuments.

The barely completed capital of the infant republic was captured and burned (1814) by the British during the War of 1812, but it was soon reconstructed. By 1860 its population was 61,100. Washington's first great period of development took place following the Civil War. The city's continuing growth, closely tied to the expansion of governmental functions, accelerated during the 1930s and particularly after World War II.

The Washington Monument was dedicated on Washington's birthday, Feb. 21, 1885. At the time it was the tallest monument of masonry in the world, standing 410 feet high. The ceremony of dedication on that cold, clear morning in 1885 was presided over by the Masonic

The unfinished obelisk lying in its quarry at Aswan.

274

The unfinished obelisk lying in its quarry at Aswan.

Grand Master of the Grand Lodge of Massachusetts who had a golden urn containing a lock of General Washington's hair.

The Mystery of the Unfinished Obelisk of Aswan

Aswan is the capital of Aswan governorate in southeastern Egypt. About 708 km (440 mi) southeast of Cairo, it is on the east bank of the Nile River about 9.6 km (6 mi) north of the First Cataract. The population swelled as a result of the construction (1960-70) of the Aswan High Dam. About 13 km (8 mi) south of the city, the dam has spurred development of metal, artificial fertilizer, and hydroelectric power industries. The older Aswan Dam is located between the city and the Aswan High Dam.

The Aswan High Dam blocks the Nile River near the resort town of Aswan in Upper Egypt. One of the world's largest structures, the rock-fill dam, completed in 1970, has a volume about 17 times that of the Great Pyramid at Giza. It is 3.26 km (2.3 mi) in length and rises 111 m (364 ft) above the riverbed. Lake Nasser (High Dam Lake), the reservoir it impounds, averages 9.6 km (6 mi) wide and extends upstream 499 km (310 mi). About 30 percent of its length is in neighboring Sudan. An earlier granite dam, the Aswan Dam, lies 6.4 km (4 mi) downstream—about midway between the Aswan High Dam and the town. The Aswan Dam was completed in 1902, but its crest has been twice raised.

Ten years in construction, the Aswan High Dam cost $1 billion. The water it stores has opened the way to agricultural expansion. More than 360,000 ha (900,000 acres), most of it formerly desert, were added to the total of arable land; an equal amount was irrigated year round to enable it to produce several crops a year instead of just one. Between 1979 and the mid-1980s, however, overuse and drought led to a 20% drop in the water level of Lake Nasser, forcing drastic reductions in the flow of irrigation water and reducing power output by 55%. The dam has a hydroelectric power capacity of 2.1 million kW and supplies more than 25% of Egypt's power.

Is it possible that an ancient dam at the first cataract of the Nile also supplied power to an ancient power plant in the Aswan vicinity? Such a power plant, hydroelectric, with copper windings and cables, much as today's plants, could have existed circa 12,000 BC. A complete hydroelectric system may have been created—and is now gone. Metals, unless they are quickly looted (which is usually the case) will rust and oxidize within only few hundred years. During this time modern-type quarrying was taking place (theoretically) at the quarries in Aswan. Giant obelisks were quarried, including the cracked "unfinished obelisk."

Also quarried would have been the stone blocks for the Osirion at Abydos. This megalithic site, half underwater in a swamp, is thought by many Egyptologists to be over 10,000 years old and is built out of perfectly cut blocks of Aswan granite weighing 100 tons or more.

At the edge of Aswan's northern granite quarries, separated from the bedrock, but lying in place, is what would have been the largest obelisk in Egypt. Because it cracked before the quarriers could lift it from its place and transport it to the Nile, we are able to follow the details of the ancient quarrier's art.

Says John Anthony West, the Egyptologist author of the books *Serpent In the Sky* and *A Traveller's Key to Ancient Egypt*, concerning the unfinished obe-

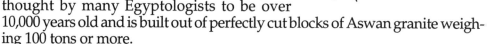

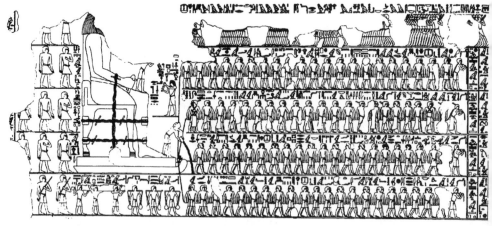

Dragging a giant statue on a sled. Were obelisks also moved this way?

lisk, "If completed, this obelisk would have been a single monolith of granite 137 feet (43 m) long and 133/4 feet (4.3 m) thick at its thick end. It would have weighed 1,168 tons. The labor involved seems almost unimaginable. Apart from a few astronaut enthusiasts, everyone agrees that the Egyptians achieved their impressive results with the simplest possible means. Rock was quarried out of its bed by drilling a series of holes. Wooden stakes were inserted into the holes and soaked with water, and the expansion of the wood cracked the block out of the bedrock along the prescribed lines. The rough blocks were then pounded smooth with balls of dolerite, a rock even harder than the granite. A number of these pounding balls have been found. Assuming they were not brought down in spaceships by the alleged astronauts, the working-up of the dolerite into pounding balls then poses its own problem, and there are no definitive solutions. In all likelihood, the Egyptians had some simple yet sophisticated method of working with extremely hard abrasives—carborundum or even ground gems. The Egyptians had no steel, and the rare iron (probably from meteorites) had a ritual not a practical purpose. The Egyptians could, however, temper copper to a hardness close to that of steel by some method we also have not been able to reproduce. It is thought likely that

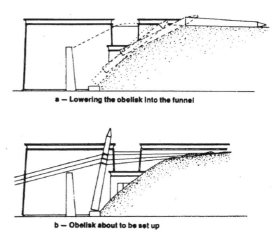

a — Lowering the obelisk into the funnel

b — Obelisk about to be set up

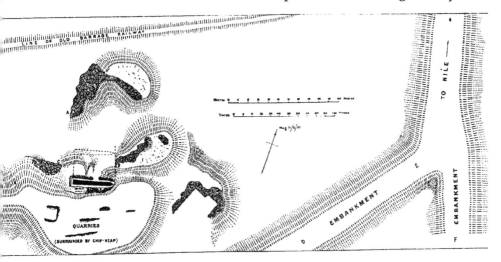

The Aswan Quarry with the Unfinished Obelisk.

they set their copper saws and drills with teeth and bits of precious gems, but there is little concrete evidence."

The "owner" of the obelisk is not known, nor where it might have been intended to go. The obelisk cracked from unknown causes, perhaps along a faultline invisible from the surface, in the final stage of freeing it from the bedrock.

Says West, "Trenches had been cut all around the monolith, and the next step would have been undercutting it, and propping it up as work progressed. At that point either the vast rough piece of stone would have been levered somehow out of its trench, or the entire Nileside wall of granite that now encloses it would have been cut away, and it would have been ready to begin its several mile trek to the river, and the long ride to its destination, which remains unknown, as is the pharaoh responsible for ordering it. It is thought that it may have been cut for Hatshepsut."

So we find that some of the giant crystals of Atlantis may still be around, used over and over. Some are still lying unfinished in their ancient quarry—a quarry that may be over 10,000 years old. They were already standing, like the Osirion, the Sphinx, and the Giza Pyramids, or they were moved and erected again by the dynastic Egyptians, often to be moved and erected again by the Romans.

As the original purpose of these towers became lost, worship of their

Moving a statue with a sled and man power.

278

"power" began. When possible, they were moved to new temples being built or them. Otherwise, they were just too huge to move and erect. They were, I believe, left-over crystal towers from "Atlantis."

One of the ancient obelisks at Axum.

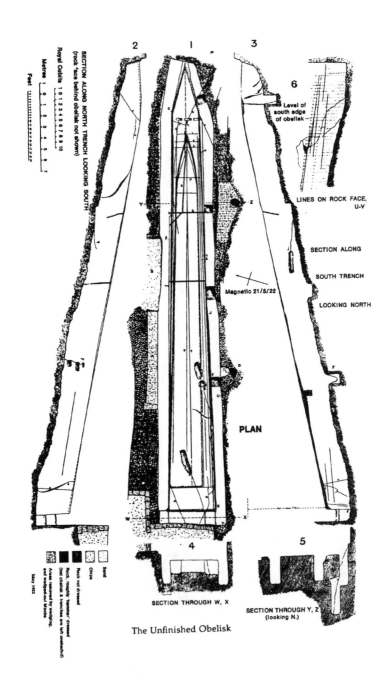

2   1   3

SECTION ALONG NORTH TRENCH LOOKING SOUTH
(rock 'face behind obelisk not shown)

Royal Cubits 1 0 1 2 3 4 5 6 7 8 9 10

Metres 1 0 1 2 3 4 5 6 7

Feet

6

Level of
south edge
of obelisk

LINES ON ROCK FACE,
U-V

Magnetic 21/5/22

SECTION ALONG

SOUTH TRENCH

LOOKING NORTH

PLAN

4   5

Sand

Chips

Rock, not dressed

Rock, roughly 'hammer' dressed
(but obelisk & trenches are left unfinished)

Areas removed by wedging,
and wedge-cut blocks

May 1922

SECTION THROUGH W, X

SECTION THROUGH Y, Z
(looking N.)

The Unfinished Obelisk

280

) *The Original Position of the Obelisk.* The New York *Herald,* 13 February 1880, carried the following: "The obelisk and its foundations will be removed and replaced in New York exactly in the positions in which they were found, each piece having been numbered to correspond with numbers on a drawing that was made before the pieces were removed." The monument was erected by Thothmes III at the outer porch of the Temple of Amen at Heliopolis, where it and its twin (now in London) guarded the entrance of the temple for 2,000 years before they were moved to Alexandria. On 12 June 1880, with the assistance of Mr. Zola, Most Worshipful Grand Master of Egypt, the obelisk was entrusted to Lieutenant Commander H. H. Gorringe, U.S.N., a member of Anglo-Saxon Lodge No. 137, for shipment to New York. On 9 October 1880 the obelisk was raised with great Masonic ceremonies in Central Park, New York City. With 8,000 Masons in attendance, the cornerstone of the ancient obelisk was laid by Jesse B. Anthony, Grand Master of Masons in New York State, as 30,000 awe-struck spectators and curiosity-seekers watched, wondering what the strange rites they were seeing performed meant. See *New York World,* 9 October 1880.

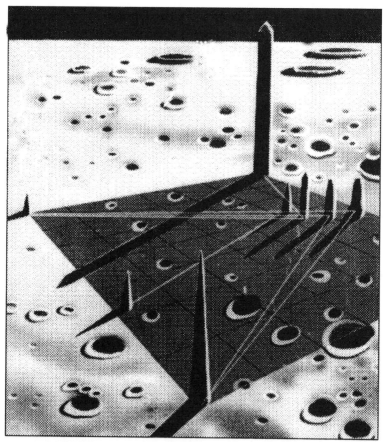

Obelisks On the Moon: Incredibly, there exists evidence for a series of obelisks in a special alignment on our Moon. The Soviet space probe Luna-9 took some startling photographs (February 4, 1966) after the vehicle had landed on the Ocean of Storms, one of those dark, circular "seas" of lava on the Earth side of the Moon. The photos revealed strange towering structures that appear to be lined up rather than scattered randomly across the lunar surface.

Dr. Ivan Sanderson, the late director of the Society for the Investigation of the Unexplained and science editor for Argosy magazine, observed that the Soviet photographs "reveal two straight lines of equidistant stones that look like markers along an airport runway. These circular stones are all identical, and are positioned at an angle that produces a strong reflection from the Sun, which would render them visible to descending aircraft." (*Argosy*, August 1970.)

But Sanderson was not the only reporter revealing these strange structures to the world. The Soviet press also carried articles on them. The Soviet magazine Technology of Youth gave an extensive report on them, calling them "stone markers" which were unquestionably "planned structures," and suggested that these "pointed pyramids" were not natural formations but definitely artificial structures of alien origin.

After examining the photographs of these objects, Dr. S. Ivanov, winner of the Laureate State Prize (which the Soviets consider equivalent to the Novel Prize), calculated from the shadows cast by the spire-like structures that at least one was about fifteen stories high.

Ivanov, who is also the inventor of stereo movies in the Soviet Union, pointed out that by luck—perhaps the space probe landed on a spot where the ground had settled, or set down upon a small stone or rough spot—"a chance displacement of Luna-9 on its horizontal axis had caused the stones to be taken at slightly different angles." This double set of photographs allowed him to produce a three-dimensional stereoscopic view of the lunar "runway."

The result of this bit of good fortune, as Ivanov reports, was that the stereoscopic effect enabled scientists to figure the distances between the spires. They found, much to their surprise, that they were spaced at regular intervals.

BIBLIOGRAPHY
& FOOTNOTES

1. *Flying Saucers Have Landed,* Desmond Leslie and George Adamski, 1953, The British Book Centre, New York.
2. *Sensitive Chaos,* Theodore Schwenk, 1965, Rudolph Steiner Press, London.
3. *Vimana In Ancient India,* Dileep Kumar Kanjilal, 1985, Sanskrit Pustak Bhandar, Calcutta.
4. *Atlantis, The Antidiluvian World,* Ignatius Donnelly, 1882, Harper & Row, NYC.
5. *War In Ancient India,* Ramachandra Dikshitar, 1944, Motilal Banarsidass, Delhi.
6. *The Bhagavad Gita,* translated by Juan Mascaro, 1962, Penguin Books, NYC.
7. *The Book of the Damned,* Charles Fort, 1919, Ace Books, NYC.
8. *The Children of Mu,* James Churchward, 1931, Ives Washburn Inc, NYC.
9. *Doomsday 1999 A.D.,* Charles Berlitz, 1981, Doubleday & Co., Garden City, NJ.
10. *Fate Magazine,* Sept. 1983, Highland Park, IL.
11. *Forgotten Worlds,* Robert Charroux, 1971, Popular Library, NYC.
12. *Lost Cities & Ancient Mysteries of South America,* David Hatcher Childress, 1985, Adventures Unlimited Press, Kempton, Illinois.
13. *Gods of Air and Darkness,* Richard Mooney, 1975, Stein & Day, NYC.
14. *The Gods Unknown,* Robert Charroux, 1969, Berkley Books, NYC.
15. *Van Nostrand's Scientific Encyclopedia,* Douglas Considine, Ed., 1983, Van Nostrand Reinhold Co New York.
16. *Mercury, UFO Messenger of the Gods,* William Clendenon, 1990, Adventure Survival Productions, Biloxi, Mississippi.
17. *Invisible Residents,* Ivan T. Sanderson, 1970, World Publishing, Cleveland, Ohio.
18. *Inside the Spaceships,* George Adamski, 1955, Adelard-Schuman, New York City.
19. *Stranger Than Science,* Frank Edwards, 1959, Lyle Stuart, NYC.
20. *Legacy of the Gods,* Robert Charroux, 1965, Robert Laffont Inc., NYC.
21. *Gods and Spacemen in the Ancient East,* W. Raymond Drake, 1968, Sphere Books, London.
22. *Gods, Demons, and Others,* R.K. Narayan, 1964, Bantam Classic Editions, Bantam Books, NY.
23. *Exploration Fawcett,* Brian Fawcett, 1953, Hutchinson & Co. London.
24. *The Fate of Colonel Fawcett,* Geraldine Cummings, 1955, Aquarian Press, London.
25. *Dimensions,* Jacques Vallee, 1988, Ballantine Books, NY.
26. *The Mahabharata,* translated by Protap Chandra Roy, 1889, Calcutta.
27. *Masters of the World,* Robert Charroux, 1967, Berkeley Books, NYC.
28. *Mysteries of Time & Space,* Brad Steiger, 1974, Prentice Hall, NYC.
29. *The Morning of the Magicians,* Louis Pauwels & Jacques Bergier, 1960, Stein & Day, NYC.
30. *Quest For Zero-Point Energy,* Moray B. King, 2001, Adventures Unlimited Press, Kempton, IL.
31. *Not Of This World,* Peter Kolosimo, 1971, University Books, Seacaucus, NJ.
32. *Lost Cities of China, Central Asia & India,* David Hatcher Childress, 1987, AUP, Stelle, Illinois.
33. *One Hundred Thousand Years of Man's Unknown History,* Robert Charroux, 1965, Robert Laffont Inc., NYC.
34. *Lost Cities of Ancient Lemuria & the Pacific,* David Hatcher Childress, 1988, AUP, Kempton, IL.
35. *The Ramayana,* R.C. Prasad, 1988, Motilal Banarsidass, Delhi.
36. *The Traveler's Key to Ancient Egypt,* John Anthony West, 1985, Alfred Knopf, NYC
37. *The Serpent in the Sky,* John Anthony West, , Harper & Row, NYC.
38. *Shambala,* Andrew Tomas, 1977, Sphere Books, London.
39. *Mysteries of Ancient South America,* 1947, 1998, Adventures Unlimited Press, Kempton, IL.
40. *Tao Te Ching, Lao Tzu,* various translations and editions.
41. *Vimana Aircraft of Ancient India & Atlantis,* D.H. Childress, ed. 1992, AUP, Kempton, Illinois.

42. *Cleopatra's Needles and Other Egyptian Obelisks*, Budge, E. A. T. W., 1926, Dover, New York.
43. *Timeless Earth*, Peter Kolosimo, 1973, Bantam Books, NYC.
44. *The Magic of Obelisks*, Peter Tompkins, 1981, Harper & Row, New York.
45. *The Obelisks of Egypt*, Labib Habachi, 1984, American University in Cairo Press, Cairo. Egypt.
46. *The Pyramids of Ancient Egypt*, Zahi A. Hawass, 1990, The Carnegie Museum of Natural History, Pittsburgh, PA.
47. *The Giza Power Plant*, Christopher Dunn, 1998, Bear & Co. Rochester, VT.
48. *Vymaanika-Shaastra Aeronautics*, Maharishi Bharadwaaja, translated and published 1973 by G.R Josyer, Mysore, India.
49. *Masers and Lasers*, H. Arthur Klein, 1963, J.B. Lippincott Co., Philadelphia & New York.
50. *Tapping the Zero-Point Energy*, Moray B. King, 1989, 2002, Adventures Unlimited Press, Kempton, IL.
51. *One Foot In Atlantis*, William Henry, 1998, Earthpulse Press, Anckorage, AK.
52. *The Giza Death Star*, Joseph Farrell, 2001, Adventures Unlimited Press, Kempton, IL.
53. *Ancient Egyptian Mysticism & Its Relevance Today*, John Van Auken, 1999, ARE Press, Virginia Beach, VA.
54. *Edgar Cayce Readings, Vol. 22: Atlantis*, 1987, Edgar Cayce Foundation, Virginia Beach, VA.
55. *Out of the Dust*, J.O. Kinnaman, Kinnaman Foundation, 1955, Longbeach, CA.
56. *The Land of Osiris*, Stephen Mehler, 2001, Adventures Unlimited Press, Kempton, IL.
57. *Mystery Religions of the Ancient World*, Joscelyn Godwin, 1981, Thames & Hudson, London.
58. *Remarkable Luminous Phenomena In Nature*, Willaim Corliss, 2001, Sourcebook Project, Glen Arm, MD.
59. *The Cosmic Conspiracy*, Stan Deyo, 1978, 1998, Deyo Enterprises, Pueblo, Colorado.
60. *The Great Pyramid: A Miracle in Stone*, Dr. Joseph Seiss, 1877, Philadelphia.
61. *Ancient Ethiopia*, David W. Philllipson, 1988, British Museum Press, London.
62.. *Travels In Ethiopia*, David Buxton, 1949, Ernest Benn Ltd., London.
63. *Ancient Egyptian Metallurgy*, H. Garland and C.O. Bannister, 1927, London.
64. *Anti-Gravity & the World Grid*, D.H. Childress, 1987, Adventures Unlimited Press, Stelle, Illinois.
65. *Edgar Cayce On Atlantis*, Hugh Lynn Cayce, 1968, A.R.E., Virginia Beach, VA.
66. *Anti-Gravity & the Unified Field*, D.H. Childress, ed. 1990, AUP, Stelle, Illinois.
67. *The Tesla Papers*, D.H. Childress, ed. 1999, AUP, Kempton, Illinois.
68. *The Fantastic Inventions of Nikola Tesla*, D.H. Childress, ed. 1993, AUP, Kempton, Illinois.
69. *Quest For Zero Point Energy*, Moray B. King, 2002, AUP, Kempton, Illinois.
70. *The Sun Rises*, Dr. Robert D. Stelle, 1952, Lemurian Fellowship, Ramona, CA.
71. *A Dweller On Two Planets*, Phylos the Thibetan, 1884, Borden Publishing, Alhambra, California.
72. *An Earth Dweller's Return*, Phylos the Thibetan, 1940, Lemurian Fellowship, Ramona, CA
73..*The Riddle of the Pyramids*, Kurt Mendelssohn, Thames & Hudson, London.
74. *Lost Science*, Gerry Vassilatos, 1998, AUP, Kempton, Illinois.
75..*Secrets of Cold War Technology*, Gerry Vassilatos, 1996, AUP, Kempton, Illinois.
76..*Technology of the Gods*, D.H. Childress, 2000, AUP, Kempton, Illinois.
77. *Easy Journey to Other Planets*, Swami Prabhupada, 1970, Bhaktivedanta Book Trust (ISKCON), Los Angeles.
78. *Irish Druids and Old Irish Religions*, James Bonwick, 1894, reprinted by Dorset Press, 1986, New York.
79. *Round Towers of Ireland*, Henry O'Brien, 1834, reprinted as *Round Towers of Atlantis*, 2002, by Adventures Unlimited Press.
80. *The Secret Life of Plants*, Christopher Bird and Peter Thompkins, 1984, Harper & Row, New York.
81. *Ancient Mysteries, Modern Visions*, Philip S. Callahan, 1984, Acres USA, Kansas City, MO.

NEW BOOKS

THE LAND OF OSIRIS
An Introduction to Khemitology
by Stephen S. Mehler

Was there an advanced prehistoric civilization in ancient Egypt? Were they the people who built the great pyramids and carved the Great Sphinx? Did the pyramids serve as energy devices and not as tombs for kings? Independent Egyptologist Stephen S. Mehler has spent over 30 years researching the answers to these questions and believes the answers are yes! Mehler has uncovered an indigenous oral tradition that still exists in Egypt, and has been fortunate to have studied with a living master of this tradition, Abd'El Hakim Awyan. Mehler has also been given permission to present these teachings to the Western world, teachings that unfold a whole new understanding of ancient Egypt and have only been presented heretofore in fragments by other researchers. Chapters include: Egyptology and Its Paradigms; Khemitology—New Paradigms; Asgat Nefer—The Harmony of Water; Khemit and the Myth of Atlantis; The Extraterrestrial Question; 17 chapters in all.

272 PAGES. 6X9 PAPERBACK. ILLUSTRATED. COLOR SECTION. BIBLIOGRAPHY. $18.95. CODE: LOOS

QUEST FOR ZERO-POINT ENERGY
Engineering Principles for "Free Energy"
by Moray B. King

King expands, with diagrams, on how free energy and anti-gravity are possible. The theories of zero point energy maintain there are tremendous fluctuations of electrical field energy embedded within the fabric of space. King explains the following topics: Tapping the Zero-Point Energy as an Energy Source; Fundamentals of a Zero-Point Energy Technology; Vacuum Energy Vortices; The Super Tube; Charge Clusters: The Basis of Zero-Point Energy Inventions; Vortex Filaments, Torsion Fields and the Zero-Point Energy; Transforming the Planet with a Zero-Point Energy Experiment; Dual Vortex Forms: The Key to a Large Zero-Point Energy Coherence. Packed with diagrams, patents and photos. With power shortages now a daily reality in many parts of the world, this book offers a fresh approach very rarely mentioned in the mainstream media.

224 PAGES. 6X9 PAPERBACK. ILLUSTRATED. $14.95. CODE: QZPE

ATLANTIS & THE POWER SYSTEM OF THE GODS
Mercury Vortex Generators & the Power System of Atlantis
by David Hatcher Childress and Bill Clendenon

Atlantis and the Power System of the Gods starts with a reprinting of the rare 1990 book Mercury: UFO Messenger of the Gods by Bill Clendenon. Clendenon takes us on an unusual voyage into the world of ancient flying vehicles, strange personal UFO sightings, a meeting with a "Man In Black" and then to a centuries-old library in India where he got his ideas for the diagrams of mercury vortex engines. The second part of the book is Childress' fascinating analysis of Nikola Tesla's broadcast system in light of Edgar Cayce's "Terrible Crystal" and the obelisks of ancient Egypt and Ethiopia. Includes: Atlantis and its crystal power towers that broadcast energy; how these incredible power stations may still exist today; inventor Nikola Tesla's nearly identical system of power transmission; Mercury Proton Gyros and mercury vortex propulsion; more. Richly illustrated, and packed with evidence that Atlantis not only existed—it had a world-wide energy system more sophisticated than ours today.

246 PAGES. 6X9 PAPERBACK. ILLUSTRATED. $15.95. CODE: APSG

THE GIZA DEATH STAR
The Paleophysics of the Great Pyramid & the Military Complex at Giza
by Joseph P. Farrell

Physicist Joseph Farrell's amazing book on the secrets of Great Pyramid of Giza. The Giza Death Star starts where British engineer Christopher Dunn leaves off in his 1998 book, The Giza Power Plant. Was the Giza complex part of a military installation over 10,000 years ago? Chapters include: An Archaeology of Mass Destruction, Thoth and Theories; The Machine Hypothesis; Pythagoras, Plato, Planck, and the Pyramid; The Weapon Hypothesis; Encoded Harmonics of the Planck Units in the Great Pyramid; High Freqquency Direct Current "Impulse" Technology; The Grand Gallery and its Crystals: Gravito-acoustic Resonators; The Other Two Large Pyramids; the "Causeways," and the "Temples"; A Phase Conjugate Howitzer; Evidence of the Use of Weapons of Mass Destruction in Ancient Times; more.

290 PAGES. 6X9 PAPERBACK. ILLUSTRATED. $16.95. CODE: GDS

THE ORION PROPHECY
Egyptian & Mayan Prophecies on the Cataclysm of 2012
by Patrick Geryl and Gino Ratinckx

In the year 2012 the Earth awaits a super catastrophe: its magnetic field reverse in one go. Phenomenal earthquakes and tidal waves will completely destroy our civilization. Europe and North America will shift thousands of kilometers northwards into polar climes. Nearly everyone will perish in the apocalyptic happenings. These dire predictions stem from the Mayans and Egyptians—descendants of the legendary Atlantis. The Atlanteans had highly evolved astronomical knowledge and were able to exactly predict the previous world-wide flood in 9792 BC. They built tens of thousands of boats and escaped to South America and Egypt. In the year 2012 Venus, Orion and several others stars will take the same 'code-positions' as in 9792 BC! For thousands of years historical sources have told of a forgotten time capsule of ancient wisdom located in a mythical labyrinth of secret chambers filled with artifacts and documents from the previous flood. We desperately need this information now—and this book gives one possible location.

324 PAGES. 6X9 PAPERBACK. ILLUSTRATED. BIBLIOGRAPHY. $16.95. CODE: ORP

ALTAI-HIMALAYA
A Travel Diary
by Nicholas Roerich

Nicholas Roerich's classic 1929 mystic travel book is back in print in this deluxe paperback edition. The famous Russian-American explorer's expedition through Sinkiang, Altai-Mongolia and Tibet from 1924 to 1928 is chronicled in 12 chapters and reproductions of Roerich's inspiring paintings. Roerich's "Travel Diary" style incorporates various mysteries and mystical arts of Central Asia including such arcane topics as the hidden city of Shambala, Agartha, more. Roerich is recognized as one of the great artists of this century and the book is richly illustrated with his original drawings.

407 PAGES. 6X9 PAPERBACK. ILLUSTRATED. $18.95. CODE: AHIM

LOST CITIES

TECHNOLOGY OF THE GODS
The Incredible Sciences of the Ancients
by David Hatcher Childress

Popular *Lost Cities* author David Hatcher Childress takes us into the amazing world of ancient technology, from computers in antiquity to the "flying machines of the gods." Childress looks at the technology that was allegedly used in Atlantis and the theory that the Great Pyramid of Egypt was originally a gigantic power station. He examines tales of ancient flight and the technology that it involved; how the ancients used electricity; megalithic building techniques; the use of crystal lenses and the fire from the gods; evidence of various high tech weapons in the past, including atomic weapons; ancient metallurgy and heavy machinery; the role of modern inventors such as Nikola Tesla in bringing ancient technology back into modern use; impossible artifacts; and more.

356 PAGES. 6X9 PAPERBACK. ILLUSTRATED. BIBLIOGRAPHY. $16.95. CODE: TGOD

VIMANA AIRCRAFT OF ANCIENT INDIA & ATLANTIS
by David Hatcher Childress, introduction by Ivan T. Sanderson

Did the ancients have the technology of flight? In this incredible volume on ancient India, authentic Indian texts such as the *Ramayana* and the *Mahabharata* are used to prove that ancient aircraft were in use more than four thousand years ago. Included in this book is the entire Fourth Century BC manuscript *Vimaanika Shastra* by the ancient author Maharishi Bharadwaaja, translated into English by the Mysore Sanskrit professor G.R. Josyer. Also included are chapters on Atlantean technology, the incredible Rama Empire of India and the devastating wars that destroyed it. Also an entire chapter on mercury vortex propulsion and mercury gyros, the power source described in the ancient Indian texts. Not to be missed by those interested in ancient civilizations or the UFO enigma.

334 PAGES. 6X9 PAPERBACK. ILLUSTRATED. $15.95. CODE: VAA

LOST CONTINENTS & THE HOLLOW EARTH
I Remember Lemuria and the Shaver Mystery
by David Hatcher Childress & Richard Shaver

Lost Continents & the Hollow Earth is Childress's thorough examination of the early hollow earth stories of Richard Shaver and the fascination that fringe fantasy subjects such as lost continents and the hollow earth have had for the American public. Shaver's rare 1948 book *I Remember Lemuria* is reprinted in its entirety, and the book is packed with illustrations from Ray Palmer's *Amazing Stories* magazine of the 1940s. Palmer and Shaver told of tunnels running through the earth—tunnels inhabited by the Deros and Teros, humanoids from an ancient spacefaring race that had inhabited the earth, eventually going underground, hundreds of thousands of years ago. Childress discusses the famous hollow earth books and delves deep into whatever reality may be behind the stories of tunnels in the earth. Operation High Jump to Antarctica in 1947 and Admiral Byrd's bizarre statements, tunnel systems in South America and Tibet, the underground world of Agartha, the belief of UFOs coming from the South Pole, more.

344 PAGES. 6X9 PAPERBACK. ILLUSTRATED. $16.95. CODE: LCHE

LOST CITIES OF NORTH & CENTRAL AMERICA
by David Hatcher Childress

Down the back roads from coast to coast, maverick archaeologist and adventurer David Hatcher Childress goes deep into unknown America. With this incredible book, you will search for lost Mayan cities and books of gold, discover an ancient canal system in Arizona, climb gigantic pyramids in the Midwest, explore megalithic monuments in New England, and join the astonishing quest for lost cities throughout North America. From the war-torn jungles of Guatemala, Nicaragua and Honduras to the deserts, mountains and fields of Mexico, Canada, and the U.S.A., Childress takes the reader in search of sunken ruins, Viking forts, strange tunnel systems, living dinosaurs, early Chinese explorers, and fantastic lost treasure. Packed with both early and current maps, photos and illustrations.

590 PAGES. 6X9 PAPERBACK. ILLUSTRATED. FOOTNOTES & BIBLIOGRAPHY. $16.95. CODE: NCA

LOST CITIES & ANCIENT MYSTERIES OF SOUTH AMERICA
by David Hatcher Childress

Rogue adventurer and maverick archaeologist David Hatcher Childress takes the reader on unforgettable journeys deep into deadly jungles, high up on windswept mountains and across scorching deserts in search of lost civilizations and ancient mysteries. Travel with David and explore stone cities high in mountain forests and hear fantastic tales of Inca treasure, living dinosaurs, and a mysterious tunnel system. Whether he is hopping freight trains, searching for secret cities, or just dealing with the daily problems of food, money, and romance, the author keeps the reader spellbound. Includes both early and current maps, photos, and illustrations, and plenty of advice for the explorer planning his or her own journey of discovery.

381 PAGES. 6X9 PAPERBACK. ILLUSTRATED. FOOTNOTES. BIBLIOGRAPHY. $14.95. CODE: SAM

LOST CITIES & ANCIENT MYSTERIES OF AFRICA & ARABIA
by David Hatcher Childress

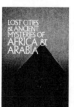

Across ancient deserts, dusty plains and steaming jungles, maverick archaeologist David Childress continues his world-wide quest for lost cities and ancient mysteries. Join him as he discovers forbidden cities in the Empty Quarter of Arabia; "Atlantean" ruins in Egypt and the Kalahari desert; a mysterious, ancient empire in the Sahara; and more. This is the tale of an extraordinary life on the road: across war-torn countries, Childress searches for King Solomon's Mines, living dinosaurs, the Ark of the Covenant and the solutions to some of the fantastic mysteries of the past.

423 PAGES. 6X9 PAPERBACK. ILLUSTRATED. FOOTNOTES & BIBLIOGRAPHY. $14.95. CODE: AFA

24 hour credit card orders—call: 815-253-6390 fax: 815-253-6300

email: auphq@frontiernet.net   www.adventuresunlimitedpress.com   www.wexclub.com

LOST CITIES

LOST CITIES OF ATLANTIS, ANCIENT EUROPE & THE MEDITERRANEAN
by David Hatcher Childress

Atlantis! The legendary lost continent comes under the close scrutiny of maverick archaeologist David Hatcher Childress in this sixth book in the internationally popular *Lost Cities* series. Childress takes the reader in search of sunken cities in the Mediterranean; across the Atlas Mountains in search of Atlantean ruins; to remote islands in search of megalithic ruins; to meet living legends and secret societies. From Ireland to Turkey, Morocco to Eastern Europe, and around the remote islands of the Mediterranean and Atlantic, Childress takes the reader on an astonishing quest for mankind's past. Ancient technology, cataclysms, megalithic construction, lost civilizations and devastating wars of the past are all explored in this book. Childress challenges the skeptics and proves that great civilizations not only existed in the past, but the modern world and its problems are reflections of the ancient world of Atlantis.
524 PAGES. 6X9 PAPERBACK. ILLUSTRATED WITH 100S OF MAPS, PHOTOS AND DIAGRAMS. BIBLIOGRAPHY & INDEX. $16.95. CODE: MED

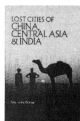

LOST CITIES OF CHINA, CENTRAL INDIA & ASIA
by David Hatcher Childress

Like a real life "Indiana Jones," maverick archaeologist David Childress takes the reader on an incredible adventure across some of the world's oldest and most remote countries in search of lost cities and ancient mysteries. Discover ancient cities in the Gobi Desert; hear fantastic tales of lost continents, vanished civilizations and secret societies bent on ruling the world; visit forgotten monasteries in forbidding snow-capped mountains with strange tunnels to mysterious subterranean cities! A unique combination of far-out exploration and practical travel advice, it will astound and delight the experienced traveler or the armchair voyager.
429 PAGES. 6X9 PAPERBACK. ILLUSTRATED. FOOTNOTES & BIBLIOGRAPHY. $14.95. CODE: CHI

LOST CITIES OF ANCIENT LEMURIA & THE PACIFIC
by David Hatcher Childress

Was there once a continent in the Pacific? Called Lemuria or Pacifica by geologists, Mu or Pan by the mystics, there is now ample mythological, geological and archaeological evidence to "prove" that an advanced and ancient civilization once lived in the central Pacific. Maverick archaeologist and explorer David Hatcher Childress combs the Indian Ocean, Australia and the Pacific in search of the surprising truth about mankind's past. Contains photos of the underwater city on Pohnpei; explanations on how the statues were levitated around Easter Island in a clockwise vortex movement; tales of disappearing islands; Egyptians in Australia; and more.
379 PAGES. 6X9 PAPERBACK. ILLUSTRATED. FOOTNOTES & BIBLIOGRAPHY. $14.95. CODE: LEM

ANCIENT TONGA
& the Lost City of Mu'a
by David Hatcher Childress

Lost Cities series author Childress takes us to the south sea islands of Tonga, Rarotonga, Samoa and Fiji to investigate the megalithic ruins on these beautiful islands. The great empire of the Polynesians, centered on Tonga and the ancient city of Mu'a, is revealed with old photos, drawings and maps. Chapters in this book are on the Lost City of Mu'a and its many megalithic pyramids, the Ha'amonga Trilithon and ancient Polynesian astronomy, Samoa and the search for the lost land of Havai'iki, Fiji and its wars with Tonga, Rarotonga's megalithic road, and Polynesian cosmology. Material on Egyptians in the Pacific, earth changes, the fortified moat around Mu'a, lost roads, more.
218 PAGES. 6X9 PAPERBACK. ILLUSTRATED. COLOR PHOTOS. BIBLIOGRAPHY. $15.95. CODE: TONG

ANCIENT MICRONESIA
& the Lost City of Nan Madol
by David Hatcher Childress

Micronesia, a vast archipelago of islands west of Hawaii and south of Japan, contains some of the most amazing megalithic ruins in the world. Part of our *Lost Cities* series, this volume explores the incredible conformations on various Micronesian islands, especially the fantastic and little-known ruins of Nan Madol on Pohnpei Island. The huge canal city of Nan Madol contains over 250 million tons of basalt columns over an 11 square-mile area of artificial islands. Much of the huge city is submerged, and underwater structures can be found to an estimated 80 feet. Islanders' legends claim that the basalt rocks, weighing up to 50 tons, were magically levitated into place by the powerful forefathers. Other ruins in Micronesia that are profiled include the Latte Stones of the Marianas, the menhirs of Palau, the megalithic canal city on Kosrae Island, megaliths on Guam, and more.
256 PAGES. 6X9 PAPERBACK. ILLUSTRATED. COLOR PHOTOS. BIBLIOGRAPHY. $16.95. CODE: AMIC

24 hour credit card orders—call: 815-253-6390 fax: 815-253-6300
email: auphq@frontiernet.net   www.adventuresunlimitedpress.com   www.wexclub.com

ATLANTIS REPRINT SERIES

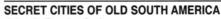

ATLANTIS: MOTHER OF EMPIRES
Atlantis Reprint Series
by Robert Stacy-Judd
Robert Stacy-Judd's classic 1939 book on Atlantis is back in print in this large-format paperback edition. Stacy-Judd was a California architect and an expert on the Mayas and their relationship to Atlantis. He was an excellent artist and his work is lavishly illustrated. The eighteen comprehensive chapters in the book are: The Mayas and the Lost Atlantis; Conjectures and Opinions; The Atlantean Theory; Cro-Magnon Man; East is West; And West is East; The Mormons and the Mayas; Astrology in Two Hemispheres; The Language of Architecture; The American Indian; Pre-Panamanians and Pre-Incas; Columns and City Planning; Comparisons and Mayan Art; The Iberian Link; The Maya Tongue; Quetzalcoatl; Summing Up the Evidence; The Mayas in Yucatan.
340 PAGES. 8X11 PAPERBACK. ILLUSTRATED. INDEX. $19.95. CODE: AMOE

MYSTERIES OF ANCIENT SOUTH AMERICA
Atlantis Reprint Series

by Harold T. Wilkins
The reprint of Wilkins' classic book on the megaliths and mysteries of South America. This book predates Wilkin's book *Secret Cities of Old South America* published in 1952. *Mysteries of Ancient South America* was first published in 1947 and is considered a classic book of its kind. With diagrams, photos and maps, Wilkins digs into old manuscripts and books to bring us some truly amazing stories of South America: a bizarre subterranean tunnel system; lost cities in the remote border jungles of Brazil; legends of Atlantis in South America; cataclysmic changes that shaped South America; and other strange stories from one of the world's great researchers. Chapters include: Our Earth's Greatest Disaster, Dead Cities of Ancient Brazil, The Jungle Light that Shines by Itself, The Missionary Men in Black: Forerunners of the Great Catastrophe, The Sign of the Sun: The World's Oldest Alphabet, Sign-Posts to the Shadow of Atlantis, The Atlanean "Subterraneans" of the Incas, Tiahuanacu and the Giants, more.
236 PAGES. 6X9 PAPERBACK. ILLUSTRATED. INDEX. $14.95. CODE: MASA

SECRET CITIES OF OLD SOUTH AMERICA
Atlantis Reprint Series
by Harold T. Wilkins
The reprint of Wilkins' classic book, first published in 1952, claiming that South America was Atlantis. Chapters include Mysteries of a Lost World; Atlantis Unveiled; Red Riddles on the Rocks; South America's Amazons Existed!; The Mystery of El Dorado and Gran Payatiti—the Final Refuge of the Incas; Monstrous Beasts of the Unexplored Swamps & Wilds; Weird Denizens of Antediluvian Forests; New Light on Atlantis from the World's Oldest Book; The Mystery of Old Man Noah and the Arks; and more.
438 PAGES. 6X9 PAPERBACK. ILLUSTRATED. BIBLIOGRAPHY & INDEX. $16.95. CODE: SCOS

THE SHADOW OF ATLANTIS
The Echoes of Atlantean Civilization Tracked through Space & Time

by Colonel Alexander Braghine
First published in 1940, *The Shadow of Atlantis* is one of the great classics of Atlantis research. The book amasses a great deal of archaeological, anthropological, historical and scientific evidence in support of a lost continent in the Atlantic Ocean. Braghine covers such diverse topics as Egyptians in Central America, the myth of Quetzalcoatl, the Basque language and its connection with Atlantis, the connections with the ancient pyramids of Mexico, Egypt and Atlantis, the sudden demise of mammoths, legends of giants and much more. Braghine was a linguist and spends part of the book tracing ancient languages to Atlantis and studying little-known inscriptions in Brazil, deluge myths and the connections between ancient languages. Braghine takes us on a fascinating journey through space and time in search of the lost continent.
288 PAGES. 6X9 PAPERBACK. ILLUSTRATED. $16.95. CODE: SOA

THE HISTORY OF ATLANTIS
by Lewis Spence
Lewis Spence's classic book on Atlantis is now back in print! Spence was a Scottish historian (1874-1955) who is best known for his volumes on world mythology and his five Atlantis books. *The History of Atlantis* (1926) is considered his finest. Spence does his scholarly best in chapters on the Sources of Atlantean History, the Geography of Atlantis, the Races of Atlantis, the Kings of Atlantis, the Religion of Atlantis, the Colonies of Atlantis, more. Sixteen chapters in all.
240 PAGES. 6X9 PAPERBACK. ILLUSTRATED WITH MAPS, PHOTOS & DIAGRAMS. $16.95. CODE: HOA

ATLANTIS IN SPAIN
A Study of the Ancient Sun Kingdoms of Spain

by E.M. Whishaw
First published by Rider & Co. of London in 1928, this classic book is a study of the megaliths of Spain, ancient writing, cyclopean walls, sun worshipping empires, hydraulic engineering, and sunken cities. An extremely rare book, it was out of print for 60 years. Learn about the Biblical Tartessus; an Atlantean city at Niebla; the Temple of Hercules and the Sun Temple of Seville; Libyans and the Copper Age; more. Profusely illustrated with photos, maps and drawings.
284 PAGES. 6X9 PAPERBACK. ILLUSTRATED. $15.95. CODE: AIS

24 hour credit card orders—call: 815-253-6390 fax: 815-253-6300
email: auphq@frontiernet.net   www.adventuresunlimitedpress.com   www.wexclub.com

TESLA TECHNOLOGY

THE FANTASTIC INVENTIONS OF NIKOLA TESLA
Nikola Tesla with additional material by David Hatcher Childress
This book is a readable compendium of patents, diagrams, photos and explanations of the many incredible inventions of the originator of the modern era of electrification. In Tesla's own words are such topics as wireless transmission of power, death rays, and radio-controlled airships. In addition, rare material on German bases in Antarctica and South America, and a secret city built at a remote jungle site in South America by one of Tesla's students, Guglielmo Marconi. Marconi's secret group claims to have built flying saucers in the 1940s and to have gone to Mars in the early 1950s! Incredible photos of these Tesla craft are included. The Ancient Atlantean system of broadcasting energy through a grid system of obelisks and pyramids is discussed, and a fascinating concept comes out of one chapter: that Egyptian engineers had to wear protective metal head-shields while in these power plants, hence the Egyptian Pharoah's head covering as well as the Face on Mars!
• His plan to transmit free electricity into the atmosphere. • How electrical devices would work using only small antennas mounted on them.
• Why unlimited power could be utilized anywhere on earth. • How radio and radar technology can be used as death-ray weapons in Star Wars. • Includes an appendix of Supreme Court documents on dismantling his free energy towers.
• Tesla's Death Rays, Ozone generators, and more…
342 PAGES. 6x9 PAPERBACK. ILLUSTRATED. BIBLIOGRAPHY AND APPENDIX. $16.95. CODE: FINT

THE TESLA PAPERS
Nikola Tesla on Free Energy & Wireless Transmission of Power
by Nikola Tesla, edited by David Hatcher Childress
In the tradition of The Fantastic Inventions of Nikola Tesla, The Anti-Gravity Handbook and The Free-Energy Device Handbook, science and UFO author David Hatcher Childress takes us into the incredible world of Nikola Tesla and his amazing inventions. Tesla's rare article "The Problem of Increasing Human Energy with Special Reference to the Harnessing of the Sun's Energy" is included. This lengthy article was originally published in the June 1900 issue of The Century Illustrated Monthly Magazine and it was the outline for Tesla's master blueprint for the world. Tesla's fantastic vision of the future, including wireless power, anti-gravity, free energy and highly advanced solar power.
Also included are some of the papers, patents and material collected on Tesla at the Colorado Springs Tesla Symposiums, including papers on:
• The Secret History of Wireless Transmission • Tesla and the Magnifying Transmitter
• Design and Construction of a half-wave Tesla Coil • Electrostatics: A Key to Free Energy
• Progress in Zero-Point Energy Research • Electromagnetic Energy from Antennas to Atoms
• Tesla's Particle Beam Technology • Fundamental Excitatory Modes of the Earth-Ionosphere Cavity
325 PAGES. 8x10 PAPERBACK. ILLUSTRATED. $16.95. CODE: TTP

LOST SCIENCE
by Gerry Vassilatos
Secrets of Cold War Technology author Vassilatos on the remarkable lives, astounding discoveries, and incredible inventions of such famous people as Nikola Tesla, Dr. Royal Rife, T.T. Brown, and T. Henry Moray. Read about the aura research of Baron Karl von Reichenbach, the wireless technology of Antonio Meucci, the controlled fusion devices of Philo Farnsworth, the earth battery of Nathan Stubblefield, and more. What were the twisted intrigues which surrounded the often deliberate attempts to stop this technology? Vassilatos claims that we are living hundreds of years behind our intended level of technology and we must recapture this "lost science."
304 PAGES. 6x9 PAPERBACK. ILLUSTRATED. BIBLIOGRAPHY. $16.95. CODE: LOS

SECRETS OF COLD WAR TECHNOLOGY
Project HAARP and Beyond
by Gerry Vassilatos
Vassilatos reveals that "Death Ray" technology has been secretly researched and developed since the turn of the century. Included are chapters on such inventors and their devices as H.C. Vion, the developer of auroral energy receivers; Dr. Selim Lemstrom's pre-Tesla experiments; the early beam weapons of Grindell-Mathews, Ulivi, Turpain and others; John Hettenger and his early beam power systems. Learn about Project Argus, Project Teak and Project Orange; EMP experiments in the 60s; why the Air Force directed the construction of a huge Ionospheric "backscatter" telemetry system across the Pacific just after WWII; why Raytheon has collected every patent relevant to HAARP over the past few years; more.
250 PAGES. 6x9 PAPERBACK. ILLUSTRATED. $15.95. CODE: SCWT

HAARP
The Ultimate Weapon of the Conspiracy
by Jerry Smith
The HAARP project in Alaska is one of the most controversial projects ever undertaken by the U.S. Government. Jerry Smith gives us the history of the HAARP project and explains how works, in technically correct yet easy to understand language. At best, HAARP is science out-of-control; at worst, HAARP could be the most dangerous device ever created, a futuristic technology that is everything from super-beam weapon to world-wide mind control device. Topics include Over-the-Horizon Radar and HAARP, Mind Control, ELF and HAARP, The Telsa Connection, The Russian Woodpecker, GWEN & HAARP, Earth Penetrating Tomography, Weather Modification, Secret Science of the Conspiracy; more. Includes the complete 1987 Eastlund patent for his pulsed super-weapon that he claims was stolen by the HAARP Project.
256 PAGES. 6x9 PAPERBACK. ILLUSTRATED. $14.95. CODE: HARP

NIKOLA TESLA'S EARTHQUAKE MACHINE
with Tesla's Original Patents
by Dale Pond and Walter Baumgartner
Now, for the first time, the secrets of Nikola Tesla's Earthquake Machine are available. Although this book discusses in detail Nikola Tesla's 1894 "Earthquake Oscillator," it is also about the new technology of sonic vibrations which produce a resonance effect that can be used to cause earthquakes. Discussed are Tesla Oscillators, Vibration Physics, Amplitude Modulated Additive Synthesis, Tele-Geo-dynamics, Solar Heat Pump Apparatus, Vortex Tube Coolers, the Serogodsky Motor, more. Plenty of technical diagrams. Be the first on your block to have a Tesla Earthquake Machine!
175 PAGES. 9x11 PAPERBACK. ILLUSTRATED. BIBLIOGRAPHY & INDEX. $16.95. CODE: TEM

Nikola Tesla's
EARTHQUAKE
MACHINE

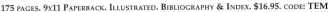

24 hour credit card orders—call: 815-253-6390 fax: 815-253-6300
email: auphq@frontiernet.net   www.adventuresunlimitedpress.com   www.wexclub.com

ANTI-GRAVITY

THE FREE-ENERGY DEVICE HANDBOOK
A Compilation of Patents and Reports
by David Hatcher Childress

A large-format compilation of various patents, papers, descriptions and diagrams concerning free-energy devices and systems. *The Free-Energy Device Handbook* is a visual tool for experimenters and researchers into magnetic motors and other "over-unity" devices. With chapters on the Adams Motor, the Hans Coler Generator, cold fusion, superconductors, "N" machines, space-energy generators, Nikola Tesla, T. Townsend Brown, and the latest in free-energy devices. Packed with photos, technical diagrams, patents and fascinating information, this book belongs on every science shelf. With energy and profit being a major political reason for fighting various wars, free-energy devices, if ever allowed to be mass distributed to consumers, could change the world! Get your copy now before the Department of Energy bans this book!
292 PAGES. 8x10 PAPERBACK. ILLUSTRATED. BIBLIOGRAPHY. $16.95. CODE: FEH

THE ANTI-GRAVITY HANDBOOK

edited by David Hatcher Childress, with Nikola Tesla, T.B. Paulicki, Bruce Cathie, Albert Einstein and others

The new expanded compilation of material on Anti-Gravity, Free Energy, Flying Saucer Propulsion, UFOs, Suppressed Technology, NASA Cover-ups and more. Highly illustrated with patents and photos. This revised and expanded edition has more material, including photos of Area 51, Nevada, the government's secret testing facility. This classic on weird science is back in a 90s format!
- **How to build a flying saucer.**
- **Arthur C. Clarke on Anti-Gravity.**
- **Crystals and their role in levitation.**
- **Secret government research and development.**
- **Nikola Tesla on how anti-gravity airships could draw power from the atmosphere.**
- **Bruce Cathie's Anti-Gravity Equation.**
- **NASA, the Moon and Anti-Gravity.**
230 PAGES. 7x10 PAPERBACK. BIBLIOGRAPHY/INDEX/APPENDIX. HIGHLY ILLUSTRATED. $14.95. CODE: AGH

ANTI–GRAVITY & THE WORLD GRID

Is the earth surrounded by an intricate electromagnetic grid network offering free energy? This compilation of material on ley lines and world power points contains chapters on the geography, mathematics, and light harmonics of the earth grid. Learn the purpose of ley lines and ancient megalithic structures located on the grid. Discover how the grid made the Philadelphia Experiment possible. Explore the Coral Castle and many other mysteries, including acoustic levitation, Tesla Shields and scalar wave weaponry. Browse through the section on anti-gravity patents, and research resources.
274 PAGES. 7x10 PAPERBACK. ILLUSTRATED. $14.95. CODE: AGW

ANTI–GRAVITY & THE UNIFIED FIELD
edited by David Hatcher Childress

Is Einstein's Unified Field Theory the answer to all of our energy problems? Explored in this compilation of material is how gravity, electricity and magnetism manifest from a unified field around us. Why artificial gravity is possible; secrets of UFO propulsion; free energy; Nikola Tesla and anti-gravity airships of the 20s and 30s; flying saucers as superconducting whirls of plasma; anti-mass generators; vortex propulsion; suppressed technology; government cover-ups; gravitational pulse drive; spacecraft & more.
240 PAGES. 7x10 PAPERBACK. ILLUSTRATED. $14.95. CODE: AGU

ETHER TECHNOLOGY
A Rational Approach to Gravity Control
by Rho Sigma

This classic book on anti-gravity and free energy is back in print and back in stock. Written by a well-known American scientist under the pseudonym of "Rho Sigma," this book delves into international efforts at gravity control and discoid craft propulsion. Before the Quantum Field, there was "Ether." This small, but informative book has chapters on John Searle and "Searle discs;" T. Townsend Brown and his work on anti-gravity and ether-vortex turbines. Includes a forward by former NASA astronaut Edgar Mitchell.
108 PAGES. 6x9 PAPERBACK. ILLUSTRATED. $12.95. CODE: ETT

TAPPING THE ZERO POINT ENERGY
Free Energy & Anti-Gravity in Today's Physics
by Moray B. King

King explains how free energy and anti-gravity are possible. The theories of the zero point energy maintain there are tremendous fluctuations of electrical field energy imbedded within the fabric of space. This book tells how, in the 1930s, inventor T. Henry Moray could produce a fifty kilowatt "free energy" machine; how an electrified plasma vortex creates anti-gravity; how the Pons/Fleischmann "cold fusion" experiment could produce tremendous heat without fusion; and how certain experiments might produce a gravitational anomaly.
170 PAGES. 5x8 PAPERBACK. ILLUSTRATED. $9.95. CODE: TAP

24 hour credit card orders—call: 815-253-6390 fax: 815-253-6300

email: auphq@frontiernet.net   www.adventuresunlimitedpress.com   www.wexclub.com

ANTI-GRAVITY

THE COSMIC MATRIX
Piece for A Jig-Saw
Part 2

Leonard G. Cramp

COSMIC MATRIX
Piece for a Jig-Saw, Part Two
by Leonard G. Cramp
Leonard G. Cramp, a British aerospace engineer, wrote his first book *Space Gravity and the Flying Saucer* in 1954. Cosmic Matrix is the long-awaited sequel to his 1966 book *UFOs & Anti-Gravity: Piece for a Jig-Saw.* Cramp has had a long history of examining UFO phenomena and has concluded that UFOs use the highest possible aeronautic science to move in the way they do. Cramp examines anti-gravity effects and theorizes that this super-science used by the craft—described in detail in the book—can lift mankind into a new level of technology, transportation and understanding of the universe. The book takes a close look at gravity control, time travel, and the interlocking web of energy between all planets in our solar system with Leonard's unique technical diagrams. A fantastic voyage into the present and future!
364 PAGES. 6x9 PAPERBACK. ILLUSTRATED. BIBLIOGRAPHY. $16.00. CODE: CMX

UFOS AND
ANTI-GRAVITY:
PIECE FOR A JIG-SAW
by Leonard G. Cramp

UFOS AND ANTI-GRAVITY
Piece For A Jig-Saw
by Leonard G. Cramp
Leonard G. Cramp's 1966 classic book on flying saucer propulsion and suppressed technology is a highly technical look at the UFO phenomena by a trained scientist. Cramp first introduces the idea of 'anti-gravity' and introduces us to the various theories of gravitation. He then examines the technology necessary to build a flying saucer and examines in great detail the technical aspects of such a craft. Cramp's book is a wealth of material and diagrams on flying saucers, anti-gravity, suppressed technology, G-fields and UFOs. Chapters include Crossroads of Aerodymanics, Aero-dynamic Saucers, Limitations of Rocketry, Gravitation and the Ether, Gravitational Spaceships, G-Field Lift Effects, The Bi-Field Theory, VTOL and Hovercraft, Analysis of UFO photos, more.
388 PAGES. 6X9 PAPERBACK. ILLUSTRATED. $16.95. CODE: UAG

THE HARMONIC CONQUEST OF SPACE
by Captain Bruce Cathie
Chapters include: Mathematics of the World Grid; the Harmonics of Hiroshima and Nagasaki; Harmonic Trans-mission and Receiving; the Link Between Human Brain Waves; the Cavity Resonance between the Earth; the Ionosphere and Gravity; Edgar Cayce—the Harmonics of the Subconscious; Stonehenge; the Harmonics of the Moon; the Pyramids of Mars; Nikola Tesla's Electric Car; the Robert Adams Pulsed Electric Motor Generator; Harmonic Clues to the Unified Field; and more. Also included are tables showing the harmonic relations between the earth's magnetic field, the speed of light, and anti-gravity/gravity acceleration at different points on the earth's surface. New chapters in this edition on the giant stone spheres of Costa Rica, Atomic Tests and Volcanic Activity, and a chapter on Ayers Rock analysed with Stone Mountain, Georgia.
248 PAGES. 6X9. PAPERBACK. ILLUSTRATED. BIBLIOGRAPHY. $16.95. CODE: HCS

THE ENERGY GRID
Harmonic 695, The Pulse of the Universe
by Captain Bruce Cathie.
This is the breakthrough book that explores the incredible potential of the Energy Grid and the Earth's Unified Field all around us. Cathie's first book, *Harmonic 33*, was published in 1968 when he was a commercial pilot in New Zealand. Since then, Captain Bruce Cathie has been the premier investigator into the amazing potential of the infinite energy that surrounds our planet every microsecond. Cathie investigates the Harmonics of Light and how the Energy Grid is created. In this amazing book are chapters on UFO Propulsion, Nikola Tesla, Unified Equations, the Mysterious Aerials, Pythagoras & the Grid, Nuclear Detonation and the Grid, Maps of the Ancients, an Australian Stonehenge examined, more.
255 PAGES. 6X9 TRADEPAPER. ILLUSTRATED. $15.95. CODE: TEG

THE BRIDGE TO INFINITY
Harmonic 371244
by Captain Bruce Cathie
Cathie has popularized the concept that the earth is crisscrossed by an electromagnetic grid system that can be used for anti-gravity, free energy, levitation and more. The book includes a new analysis of the harmonic nature of reality, acoustic levitation, pyramid power, harmonic receiver towers and UFO propulsion. It concludes that today's scien-tists have at their command a fantastic store of knowledge with which to advance the welfare of the human race.
204 PAGES. 6X9 TRADEPAPER. ILLUSTRATED. $14.95. CODE: BTF

MAN-MADE UFOS 1944—1994
Fifty Years of Suppression
by Renato Vesco & David Hatcher Childress
A comprehensive look at the early "flying saucer" technology of Nazi Germany and the genesis of man-made UFOs. This book takes us from the work of captured German scientists to escaped battalions of Germans, secret communi-ties in South America and Antarctica to todays state-of-the-art "Dreamland" flying machines. Heavily illustrated, this astonishing book blows the lid off the "government UFO conspiracy" and explains with technical diagrams the technology involved. Examined in detail are secret underground airfields and factories; German secret weapons; "suction" aircraft; the origin of NASA; gyroscopic stabilizers and engines; the secret Marconi aircraft factory in South America; and more. Introduction by W.A. Harbinson, author of the Dell novels *GENESIS* and *REVELATION*.
318 PAGES. 6X9 PAPERBACK. ILLUSTRATED. INDEX & FOOTNOTES. $18.95. CODE: MMU

24 hour credit card orders—call: 815-253-6390 fax: 815-253-6300
email: auphq@frontiernet.net   www.adventuresunlimitedpress.com   www.wexclub.com

FREE ENERGY SYSTEMS

LOST SCIENCE
by Gerry Vassilatos
Rediscover the legendary names of suppressed scientific revolution—remarkable lives, astounding discoveries, and incredible inventions which would have produced a world of wonder. How did the aura research of Baron Karl von Reichenbach prove the vitalistic theory and frighten the greatest minds of Germany? How did the physiophone and wireless of Antonio Meucci predate both Bell and Marconi by decades? How does the earth battery technology of Nathan Stubblefield portend an unsuspected energy revolution? How did the geoaetheric engines of Nikola Tesla threaten the establishment of a fuel-dependent America? The microscopes and virus-destroying ray machines of Dr. Royal Rife provided the solution for every world-threatening disease. Why did the FDA and AMA together condemn this great man to Federal Prison? The static crashes on telephone lines enabled Dr. T. Henry Moray to discover the reality of radiant space energy. Was the mysterious "Swedish stone," the powerful mineral which Dr. Moray discovered, the very first historical instance in which stellar power was recognized and secured on earth? Why did the Air Force initially fund the gravitational warp research and warp-cloaking devices of T. Townsend Brown and then reject it? When the controlled fusion devices of Philo Farnsworth achieved the "break-even" point in 1967 the FUSOR project was abruptly cancelled by ITT.
304 PAGES. 6x9 PAPERBACK. ILLUSTRATED. BIBLIOGRAPHY. $16.95. CODE: LOS

SECRETS OF COLD WAR TECHNOLOGY
Project HAARP and Beyond
by Gerry Vassilatos
Vassilatos reveals that "Death Ray" technology has been secretly researched and developed since the turn of the century. Included are chapters on such inventors and their devices as H.C. Vion, the developer of auroral energy receivers; Dr. Selim Lemström's pre-Tesla experiments; the early beam weapons of Grindell-Mathews, Ulivi, Turpain and others; John Hettenger and his early beam power systems. Learn about Project Argus, Project Teak and Project Orange; EMP experiments in the 60s; why the Air Force directed the construction of a huge Ionospheric "backscatter" telemetry system across the Pacific just after WWII; why Raytheon has collected every patent relevant to HAARP over the past few years; more.
250 PAGES. 6x9 PAPERBACK. ILLUSTRATED. $15.95. CODE: SCWT

QUEST FOR ZERO-POINT ENERGY
Engineering Principles for "Free Energy"
by Moray B. King
King expands, with diagrams, on how free energy and anti-gravity are possible. The theories of zero point energy maintain there are tremendous fluctuations of electrical field energy embedded within the fabric of space. King explains the following topics: Tapping the Zero-Point Energy as an Energy Source; Fundamentals of a Zero-Point Energy Technology; Vacuum Energy Vortices; The Super Tube; Charge Clusters: The Basis of Zero-Point Energy Inventions; Vortex Filaments, Torsion Fields and the Zero-Point Energy; Transforming the Planet with a Zero-Point Energy Experiment; Dual Vortex Forms: The Key to a Large Zero-Point Energy Coherence. Packed with diagrams, patents and photos. With power shortages now a daily reality in many parts of the world, this book offers a fresh approach very rarely mentioned in the mainstream media.
224 PAGES. 6x9 PAPERBACK. ILLUSTRATED. $14.95. CODE: QZPE

Quest For Zero Point Energy

Engineering Principles For "Free Energy"

Moray B. King

THE TIME TRAVEL HANDBOOK
A Manual of Practical Teleportation & Time Travel
edited by David Hatcher Childress
In the tradition of *The Anti-Gravity Handbook* and *The Free-Energy Device Handbook*, science and UFO author David Hatcher Childress takes us into the weird world of time travel and teleportation. Not just a whacked-out look at science fiction, this book is an authoritative chronicling of real-life time travel experiments, teleportation devices and more. *The Time Travel Handbook* takes the reader beyond the government experiments and deep into the uncharted territory of early time travellers such as Nikola Tesla and Guglielmo Marconi and their alleged time travel experiments, as well as the Wilson Brothers of EMI and their connection to the Philadelphia Experiment—the U.S. Navy's forays into invisibility, time travel, and teleportation. Childress looks into the claims of time travelling individuals, and investigates the unusual claim that the pyramids on Mars were built in the future and sent back in time. A highly visual, large format book, with patents, photos and schematics. Be the first on your block to build your own time travel device!
316 PAGES. 7X10 PAPERBACK. ILLUSTRATED. $16.95. CODE: TTH

THE TESLA PAPERS
Nikola Tesla on Free Energy & Wireless Transmission of Power
by Nikola Tesla, edited by David Hatcher Childress
David Hatcher Childress takes us into the incredible world of Nikola Tesla and his amazing inventions. Tesla's rare article "The Problem of Increasing Human Energy with Special Reference to the Harnessing of the Sun's Energy" is included. This lengthy article was originally published in the June 1900 issue of *The Century Illustrated Monthly Magazine* and it was the outline for Tesla's master blueprint for the world. Tesla's fantastic vision of the future, including wireless power, anti-gravity, free energy and highly advanced solar power. Also included are several of the papers, patents and material collected on Tesla at the Colorado Springs Tesla Symposiums, including papers on: •The Secret History of Wireless Transmission •Tesla and the Magnifying Transmitter •Design and Construction of a Half-Wave Tesla Coil •Electrostatics: A Key to Free Energy •Progress in Zero-Point Energy Research •Electromagnetic Energy from Antennas to Atoms •Tesla's Particle Beam Technology •Fundamental Excitatory Modes of the Earth-Ionosphere Cavity
325 PAGES. 8X10 PAPERBACK. ILLUSTRATED. $16.95. CODE: TTP

THE FANTASTIC INVENTIONS OF NIKOLA TESLA
by Nikola Tesla with additional material by David Hatcher Childress
This book is a readable compendium of patents, diagrams, photos and explanations of the many incredible inventions of the originator of the modern era of electrification. In Tesla's own words are such topics as wireless transmission of power, death rays, and radio-controlled airships. In addition, rare material on German bases in Antarctica and South America, and a secret city built at a remote jungle site in South America by one of Tesla's students, Guglielmo Marconi. Marconi's secret group claims to have built flying saucers in the 1940s and to have gone to Mars in the early 1950s! Incredible photos of these Tesla craft are included. The Ancient Atlantean system of broadcasting energy through a grid system of obelisks and pyramids is discussed, and a fascinating concept comes out of one chapter: that Egyptian engineers had to wear protective metal head-shields while in these power plants, hence the Egyptian Pharoah's head covering as well as the Face on Mars! •His plan to transmit free electricity into the atmosphere. •How electrical devices would work using only small antennas. •Why unlimited power could be utilized anywhere on earth. •How radio and radar technology can be used as death-ray weapons in Star Wars.
342 PAGES. 6x9 PAPERBACK. ILLUSTRATED. $16.95. CODE: FINT

24 hour credit card orders—call: 815-253-6390 fax: 815-253-6300
email: auphq@frontiernet.net   www.adventuresunlimitedpress.com   www.wexclub.com

ANTI-GRAVITY

THE A.T. FACTOR
A Scientists Encounter with UFOs: Piece For A Jigsaw Part 3
by Leonard Cramp
British aerospace engineer Cramp began much of the scientific anti-gravity and UFO propulsion analysis back in 1955 with his landmark book *Space, Gravity & the Flying Saucer* (out-of-print and rare). His next books (available from Adventures Unlimited) *UFOs & Anti-Gravity: Piece for a Jig-Saw* and *The Cosmic Matrix: Piece for a Jig-Saw Part 2* began Cramp's in depth look into gravity control, free-energy, and the interlocking web of energy that pervades the universe. In this final book, Cramp brings to a close his detailed and controversial study of UFOs and Anti-Gravity.
324 PAGES. 6x9 PAPERBACK. ILLUSTRATED. BIBLIOGRAPHY. INDEX. $16.95. CODE: ATF

COSMIC MATRIX
Piece for a Jig-Saw, Part Two
by Leonard G. Cramp

Cosmic Matrix is the long-awaited sequel to his 1966 book *UFOs & Anti-Gravity: Piece for a Jig-Saw.* Cramp has had a long history of examining UFO phenomena and has concluded that UFOs use the highest possible aeronautic science to move in the way they do. Cramp examines anti-gravity effects and theorizes that this super-science used by the craft—described in detail in the book—can lift mankind into a new level of technology, transportation and understanding of the universe. The book takes a close look at gravity control, time travel, and the interlocking web of energy between all planets in our solar system with Leonard's unique technical diagrams. A fantastic voyage into the present and future!
364 PAGES. 6x9 PAPERBACK. ILLUSTRATED. BIBLIOGRAPHY. $16.00. CODE: CMX

UFOS AND ANTI-GRAVITY
Piece For A Jig-Saw
by Leonard G. Cramp

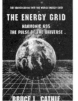

Leonard G. Cramp's 1966 classic book on flying saucer propulsion and suppressed technology is a highly technical look at the UFO phenomena by a trained scientist. Cramp first introduces the idea of 'anti-gravity' and introduces us to the various theories of gravitation. He then examines the technology necessary to build a flying saucer and examines in great detail the technical aspects of such a craft. Cramp's book is a wealth of material and diagrams on flying saucers, anti-gravity, suppressed technology, G-fields and UFOs. Chapters include Crossroads of Aerodynamics, Aerodynamic Saucers, Limitations of Rocketry, Gravitation and the Ether, Gravitational Spaceships, G-Field Lift Effects, The Bi-Field Theory, VTOL and Hovercraft, Analysis of UFO photos, more.
388 PAGES. 6x9 PAPERBACK. ILLUSTRATED. $16.95. CODE: UAG

THE ENERGY GRID
Harmonic 695, The Pulse of the Universe
by Captain Bruce Cathie.

This is the breakthrough book that explores the incredible potential of the Energy Grid and the Earth's Unified Field all around us. Cathie's first book, *Harmonic 33*, was published in 1968 when he was a commercial pilot in New Zealand. Since then, Captain Bruce Cathie has been the premier investigator into the amazing potential of the infinite energy that surrounds our planet every microsecond. Cathie investigates the Harmonics of Light and how the Energy Grid is created. In this amazing book are chapters on UFO Propulsion, Nikola Tesla, Unified Equations, the Mysterious Aerials, Pythagoras & the Grid, Nuclear Detonation and the Grid, Maps of the Ancients, an Australian Stonehenge examined, more.
255 PAGES. 6x9 TRADEPAPER. ILLUSTRATED. $15.95. CODE: TEG

THE BRIDGE TO INFINITY
Harmonic 371244
by Captain Bruce Cathie

Cathie has popularized the concept that the earth is crisscrossed by an electromagnetic grid system that can be used for anti-gravity, free energy, levitation and more. The book includes a new analysis of the harmonic nature of reality, acoustic levitation, pyramid power, harmonic receiver towers and UFO propulsion. It concludes that today's scientists have at their command a fantastic store of knowledge with which to advance the welfare of the human race.
204 PAGES. 6x9 TRADEPAPER. ILLUSTRATED. $14.95. CODE: BTF

THE HARMONIC CONQUEST OF SPACE
by Captain Bruce Cathie
Chapters include: Mathematics of the World Grid; the Harmonics of Hiroshima and Nagasaki; Harmonic Transmission and Receiving; the Link Between Human Brain Waves; the Cavity Resonance between the Earth; the Ionosphere and Gravity; Edgar Cayce—the Harmonics of the Subconscious; Stonehenge; the Harmonics of the Moon; the Pyramids of Mars; Nikola Tesla's Electric Car; the Robert Adams Pulsed Electric Motor Generator; Harmonic Clues to the Unified Field; and more. Also included are tables showing the harmonic relations between the earth's magnetic field, the speed of light, and anti-gravity/gravity acceleration at different points on the earth's surface. New chapters in this edition on the giant stone spheres of Costa Rica, Atomic Tests and Volcanic Activity, and a chapter on Ayers Rock analysed with Stone Mountain, Georgia.
248 PAGES. 6x9. PAPERBACK. ILLUSTRATED. BIBLIOGRAPHY. $16.95. CODE: HCS

MAN-MADE UFOS 1944—1994
Fifty Years of Suppression
by Renato Vesco & David Hatcher Childress
A comprehensive look at the early "flying saucer" technology of Nazi Germany and the genesis of man-made UFOs. This book takes us from the work of captured German scientists to escaped battalions of Germans, secret communities in South America and Antarctica to todays state-of-the-art "Dreamland" flying machines. Heavily illustrated, this astonishing book blows the lid off the "government UFO conspiracy" and explains with technical diagrams the technology involved. Examined in detail are secret underground airfields and factories; German secret weapons; "suction" aircraft; the origin of NASA; gyroscopic stabilizers and engines; the secret Marconi aircraft factory in South America; and more. Introduction by W.A. Harbinson, author of the Dell novels *GENESIS* and *REVELATION*.
318 PAGES. 6x9 PAPERBACK. ILLUSTRATED. INDEX & FOOTNOTES. $18.95. CODE: MMU

24 hour credit card orders—call: 815-253-6390 fax: 815-253-6300
email: auphq@frontiernet.net   www.adventuresunlimitedpress.com   www.wexclub.com

One Adventure Place
P.O. Box 74
Kempton, Illinois 60946
United States of America
Tel.: 815-253-6390 • Fax: 815-253-6300
Email: auphq@frontiernet.net
http://www.adventuresunlimitedpress.com
or www.wexclub.com/aup

ORDERING INSTRUCTIONS

✓ Remit by USD$ Check, Money Order or Credit Card
✓ Visa, Master Card, Discover & AmEx Accepted
✓ Prices May Change Without Notice
✓ 10% Discount for 3 or more Items

SHIPPING CHARGES

United States

✓ Postal Book Rate { $3.00 First Item
50¢ Each Additional Item

✓ Priority Mail { $4.00 First Item
$2.00 Each Additional Item

✓ UPS { $5.00 First Item
$1.50 Each Additional Item

NOTE: UPS Delivery Available to Mainland USA Only

Canada

✓ Postal Book Rate { $6.00 First Item
$2.00 Each Additional Item

✓ Postal Air Mail { $8.00 First Item
$2.50 Each Additional Item

✓ Personal Checks or Bank Drafts MUST BE
USD$ and Drawn on a US Bank
✓ Canadian Postal Money Orders OK
✓ Payment MUST BE USD$

All Other Countries

✓ Surface Delivery { $10.00 First Item
$4.00 Each Additional Item

✓ Postal Air Mail { $14.00 First Item
$5.00 Each Additional Item

✓ Payment MUST BE USD$
✓ Checks and Money Orders MUST BE USD$
and Drawn on a US Bank or branch.
✓ Add $5.00 for Air Mail Subscription to
Future *Adventures Unlimited* Catalogs

SPECIAL NOTES

✓ RETAILERS: Standard Discounts Available
✓ BACKORDERS: We Backorder all Out-of-
Stock Items Unless Otherwise Requested
✓ PRO FORMA INVOICES: Available on Request
✓ VIDEOS: NTSC Mode Only. Replacement only.
✓ For PAL mode videos contact our other offices:

European Office:
Adventures Unlimited, Pannewal 22,
Enkhuizen, 1602 KS, The Netherlands
http: www.adventuresunlimited.nl
Check Us Out Online at:
www.adventuresunlimitedpress.com

Please check: ☑

☐ This is my first order   ☐ I have ordered before   ☐ This is a new address

Name					
Address					
City					
State/Province			Postal Code		
Country					
Phone day		Evening			
Fax					

Item Code	Item Description	Price	Qty	Total

Please check: ☑

☐ Postal-Surface
☐ Postal-Air Mail
(Priority in USA)
☐ UPS
(Mainland USA only)

Subtotal ➠	
Less Discount-10% for 3 or more items ➠	
Balance ➠	
Illinois Residents 6.25% Sales Tax ➠	
Previous Credit ➠	
Shipping ➠	
Total (check/MO in USD$ only) ➠	

☐ Visa/MasterCard/Discover/Amex

Card Number

Expiration Date

10% Discount When You Order 3 or More Items!

Comments & Suggestions   Share Our Catalog with a Friend